Nanoscale Science and Engineering Forum 2020

Held at the 2020 AIChE Annual Meeting

Online
16 - 20 November 2020

ISBN: 978-1-7138-2315-5

Printed from e-media with permission by:

Curran Associates, Inc.
57 Morehouse Lane
Red Hook, NY 12571

Some format issues inherent in the e-media version may also appear in this print version.

Printed with permission by Curran Associates, Inc. (2021)

For permission requests, please contact AIChE
at the address below.

AIChE
120 Wall Street, FL 23
New York, NY 10005-4020

Phone: (800) 242-4363
Fax: (203) 775-5177

www.aiche.org

Additional copies of this publication are available from:

Curran Associates, Inc.
57 Morehouse Lane
Red Hook, NY 12571 USA
Phone: 845-758-0400
Fax: 845-758-2633
Email: curran@proceedings.com
Web: www.proceedings.com

Nanoscale Science and Engineering Forum 2020

Held at the 2020 AIChE Annual Meeting

Online
16 - 20 November 2020

AIChE
The Global Home of Chemical Engineers

TABLE OF CONTENTS

611723 Investigating SERS of Electrogenic Bacteria Via Charge Transfer Enhancement By Graphene Quantum Dots (GQDs)

Abstract:

We report on the enhancement of the phononic excitation in electrogenic bacteria due to the electron transfer to interfaced graphene quantum dots (GQDs). This phenomenon enables GQDs to detect otherwise unobservable bacterial Raman peaks by chemeo-phononic amplification by a factor of ten. We show confirmation of the charge transfer ability of the electrogenic bacteria into GQD and provides a platform for chemical enhancement phenomena for biological sensing at the cellular level. This study also reports on the electrogenic bacteria's phonon-polarization directly related to the chemical enhancement from the GQDs.

608557 Near-Infrared Optical Detection of Dopamine Based on Xeno Nucleic Acid (XNA) Sensors

Abstract:

Single-walled carbon nanotubes (SWCNTs) emit near-infrared fluorescence that is beneficial for in vitro and in vivo optical sensing. State-of-the-art SWCNT sensors are functionalised with single-stranded DNA (ssDNA) sequences that control the dispersivity, brightness, biocompatibility, and selectivity of the sensor. However, the fluorescence of the ssDNA-functionalized nanotube sensors is sensitive to changes in the local ion concentration. This sensitivity is problematic not only for measurements in complex biofluids, but also for sensing neurotransmitters, whose concentrations are strongly coupled to spatial and temporal ion fluctuations. The fluctuating ion concentrations can therefore interfere with the accurate optical detection of neurotransmitters, thereby limiting the use of ssDNA-SWCNT sensors for detecting and treating neurological diseases.

We present an optical SWCNT dopamine sensor with improved stability towards cation-induced fluorescence changes. The synthetic biology approach presented in this work serves as a complementary means for enhancing nanotube optoelectronic behaviour, unlocking previously unexplored possibilities for developing nano-bioengineered sensors with augmented capabilities.

608015 A Wearable Microfiber Biomaterial Incorporates Optical Nanosensors and Enables Wireless Monitoring of Oxidative Stress

Abstract:

Reactive oxygen species (ROS) are a family of signaling molecules continuously generated, transformed, and consumed by all aerobic organisms. Wound healing is one of the most dynamic biological process featuring ROS-linked cellular signaling throughout the entire process. Additionally, basal concentrations of ROS aid in the fight against invading microorganisms into open wounds. The excessive and uncontrolled production of ROS contributes to the sustaining and deregulation of inflammation processes, which play a central role in the pathogenesis of chronic non-healing wounds. Physiologically, hydrogen peroxide (referred herein as peroxide) and superoxide serve as intracellular ROS messengers stimulating key phases of wound healing including cell recruitment, production of cytokines, and angiogenesis. Of note, peroxide acts as the principal secondary messenger in wound healing and is present at low concentrations (100–250 μM) in normal wounds. Increased peroxide concentration is a biomarker for chronic wounds in which biofilm-making pathogens can grow significantly faster than normal acute wounds. Due to the short half-life of peroxide, detection and quantification of their concentrations are difficult in vivo and in patients. Thus, peroxide levels in wounds have generally been determined indirectly through analysis of downstream oxidation products of lipids, proteins, or DNA.

Single-Walled Carbon Nanotubes (SWCNTs) with engineered wrappings have recently been developed and utilized in various disparate fields ranging from additives that strengthen material composites to biomedical applications including targeted drug delivery, optical biosensing, and biological imaging. Of significant interest, SWCNTs exhibit intrinsic photoluminescence (fluorescence) that is photostable, tunable, and sensitive to its local environment. The electronic band gap energies of SWCNTs are dependent on their chiral identity and are differentially and reversibly oxidized, enabling the ratiometric detection of local redox environments.

With revolutionary advances in nanotechnology and biomaterials in the recent years, an extensive range of smart wound care biomaterials have been developed which incorporate components such as growth factors and/or antimicrobial agents such as iodine or silver. Electrospun micro/nanofibers are one of the novel classes of wound dressings as they mimic the chemical and mechanical environment of the 3D extracellular matrix. Micro/nanofiber-based wound dressings have been designed to enhance cell migration, prevent inflammation and infection and inhibit scar formation on wounds. Herein, we developed a process to construct wearable textiles incorporating the DNA-wrapped SWCNT nanosensors through a one-step co-axial electrospinning method. We then developed an optical microfibrous biomaterial with a "core-shell" morphology that encapsulates fluorescent SWCNT nanosensors. We chose Polycaprolactone (PCL) as the shell material as it is an FDA-approved polymer which has been extensively studied for tissue engineering and wound healing applications. Our optical biomaterial was able to ratiometrically detect peroxide in the wound biological range (20-500 uM). The ratiometric characteristic of the sensor will significantly facilitate the future in vivo and clinical applications as it transduces an absolute signal which is not dependent on excitation source distance and exposure time. Moreover, the optical nature of the biomaterial will enable the wireless detection without a need to disturb the wound dressing. We revealed that the signal from the biomaterial does not significantly decay over time as the fluorescence of SWCNTs is non-photobleachable. We also quantified the absolute weight of SWCNT nanosenors in the biomaterial over time using confocal -Raman microscopy and indicated that the nanosensors do not leak out of the fibers over time, mitigating any toxicological concerns related to SWCNTs. Our optical biomaterial was integrated into other common wound dressing materials such as hydrogels and commercialized bandages, showing the compatibility of this platform with other existing wound dressings. Thus, our wearable microfiber biomaterial enables wireless monitoring of oxidative stress directly from the wound site.

610131 Inverse Emulsion Synthesis of Hydrogel-Coated Gold Nanoshells for Protein Biomarker Quantification

Abstract:

As proteomics research has begun to identify increasing numbers of differences in the protein makeup of patient samples, such as saliva, tears, urine, or serum, between healthy and diseased patients, there has been growing interest in developing new biosensor diagnostic tools that allow for minimally-invasive quantification of these potential biomarkers. One strategy is differential sensor arrays that exploit semi-selective recognitive elements in order to screen for differences in multiple protein biomarkers in a single assay. However, accurate protein recognition is challenged by the similarity of the target protein's structures and properties to those of the other proteins present in the sample, necessitating the use of multiple sensing elements in order to separate the interactions. In order address this challenge, we explore the capabilities of using inverse emulsion polymerization to form libraries of multifunctional hydrogel networks on the surface of gold nanoshells (AuNS) as semi-selective recognition elements for use in a localized surface plasmon resonance-based biosensor for the detection of autoimmune disease biomarkers.

Hydrogel nanoparticles were formed by an inverse emulsion copolymerization of methyacrylic acid with acrylamide or *N*-isopropylacrylamide, crosslinked with *N,N'*-methylenebis(acrylamide), and characterized by dynamic light scattering, potentiometric titration, and Fourier-transform infrared spectroscopy. AuNS were prepared using the method of seeded growth of gold colloids on aminated silica nanoparticles followed by encapsulation with poly(maleic anhydride-*alt*-1-octadecene)-g-poly(ethylene glycol) methacrylate (PMAO-g-PEGMA) graft copolymer to improve particle stability. Hydrogel-coated AuNS were prepared using the same nanoparticle synthesis method with the addition of PMAO-g-PEGMA AuNS to the aqueous phase of the inverse emulsion. UV-Vis spectroscopy was used to measure the shifts in the wavelength of the localized surface plasmon resonance for the hydrogel-coated AuNS. The successful synthesis of coated AuNS with both formulations shows immense promise for the creation of a library of hydrogels with various functional groups that could be used to alter affinity for different protein biomarkers from patient biofluids.

Work supported by National Institutes of Health Grant R01-EB022025

606325 Imaging Healthy and Diseased Striatal Dopamine Release with Near-Infrared Catecholamine Nanosensors

Abstract:

Neuromodulators such as dopamine and serotonin and play a critical role in healthy brain function, and their disruption has been implicated in a number of neurological and psychiatric disorders. In particular, the neuromodulator dopamine is thought to facilitate learning, motivation, and motor control. Disruption of dopamine signaling plays a crucial role in a number of neurological and psychiatric disorders, including Parkinson's Disease, Huntington's Disease, and Schizophrenia. Present treatments for these disorders often broadly target dopamine neuromodulation, and many novel therapies (RNAi, cell replacement) target circuits that participate or are affected by dopaminergic systems within the brain. However, despite the established relationship between these disorders and dopamine, the exact mechanisms through which these diseases progress, and their therapies act, is unknown. This is in part because our ability to study dopaminergic modulation in healthy and diseased tissue has been limited. Unlike classical neurotransmission, dopamine exerts it influence by diffusing beyond the synaptic cleft to influence the excitability of neighboring neurons. As such, being able to study the release and spread of dopamine at the spatio-temporal resolution of single dopamine release sites (μm, ms) is critical.

Here, we implement the near-infrared fluorescent catecholamine nanosensors (nIRCats) to image dopamine release within the brain striatum of R6/2 Huntington's Disease Model (R6/2) mice. We have previously shown that these near-infrared fluorescent, polymer-functionalized semiconducting single wall carbon nanotubes serve as adept dopamine sensors in the basal ganglia, exhibiting $\Delta F/F$ of up to 24-fold in the fluorescence emission window of 1000-1300 nm, a wavelength ideal for imaging in optically scattering brain tissue[1]. Furthermore, the chemically synthetic molecular recognition elements of nIRCats readily allows imaging of dopamine dynamics in diseased tissues or tissues that have undergone treatment via novel therapy without the need for additional genetic manipulation or mouse-line development. These unique properties of nIRCat make it an ideal tool to study the dysregulation of dopamine over the course of Huntington's Disease and subsequent treatment. We demonstrate that nIRCats imaging in R6/2 Mice *ex vivo* brain slices before and after the onset of motor symptoms reveal insights into possible mechanisms of dopamine decrease with Huntington's Disease motor symptom development. These findings provide a framework to understand how the diseased brain adapts to disease progression and subsequent treatment, and underscore the utility of nIRCats as a versatile new optical tool for dopamine detection in the brain.

1. Beyene, A. G., Delevich, K. et al. "Imaging Striatal Dopamine Release Using a Non-Genetically Encoded Near-Infrared Fluorescent Catecholamine Nanosensor." *Science Advances*, **5** *(7)* (2019).

600826 Varying Oleylamine to Dibenzyl Ether Ratio for Fine-Tuning Manganese Oxide Nanoparticle Diameter and MRI Signal Intensity

Abstract:

The ultimate goal of this research project is to develop a contrast agent to reduce the number of false positive and false negative imaging results associated with current breast cancer diagnostic imaging techniques. Gadolinium chelates are the current clinically approved contrast agents for breast magnetic resonance imaging (MRI), but they are always "on" and highlight any vascularized structure. Due to their lack of targeting, both benign and malignant breast tumors are enhanced with gadolinium. By utilizing tumor-targeted pH-sensitive manganese oxide (MnO) nanoparticles, a contrast agent can be developed that will only turn "on" in the presence of the lower pH of endosomes/lysosomes inside cancer cells. Reducing the size of MnO nanoparticles should increase dissolution of MnO to Mn^{2+} at low pH to enhance MRI signal. MnO nanoparticles were synthesized through thermal decomposition of Mn(II) acetylacetonate in dibenzyl ether (DE) and oleylamine (OA) at a peak reaction temperature of 280°C for 30 minutes. The following ratios of OA:DE were used: 60:0, 50:10, 40:20, 30:30, 20:40, and 10:50 to investigate the effects of reactant conditions on nanoparticle size, Mn^{2+} release and MRI signal. MnO nanoparticles were characterized physically and chemically using transmission electron microscopy (TEM), x-ray diffraction (XRD), and Fourier-transform infrared spectroscopy (FTIR). According to TEM, the 10:50 ratio created the smallest nanoparticle diameter of 19 ± 6 nm, while the 50:10 ratio created the largest nanoparticle diameter of 39 ± 15 nm. All nanoparticles were comprised of MnO (XRD) coated with oleylamine (FTIR). To further elucidate the effect of nanoparticle size and composition on the release of Mn^{2+} and resulting MRI signal, the smallest MnO nanoparticles (19 nm) were compared with MnO nanoparticles of a larger diameter (32 nm) and Mn_3O_4 nanoparticles of similar diameter (17 nm). Nanoparticles were incubated at 3 different pHs to measure Mn^{2+} release over time including pH 7.4 (blood pH), 6.5 (pH of tumor extracellular space), and 5 (endosome/lysosome pH) using inductively coupled plasma-optical emission spectroscopy (ICP-OES) and 1T MRI. The smallest MnO nanoparticles were the most efficient MRI contrast agents by releasing the highest amount of Mn^{2+} at pH 5 (~50% at 24 hr) and 6.5 (~11% at 24 hr), which produced the lowest T_1 value (646 ms at 24 hr, pH 5) and maximum MRI signal at 1T; Mn_3O_4 nanoparticles generated the lowest MRI signal. Therefore, MnO nanoparticles instead of Mn_3O_4 should be utilized, and the nanoparticles should be targeted intracellularly to maximize MRI contrast generation. Future work includes fabricating hydrophilic MnO nanoparticles using phospholipid encapsulation and adding breast cancer-targeting peptides to enable specific detection of malignant versus benign breast tumors.

597260 Understanding Corona Exchange Dynamics on Carbon Nanotubes with Multiplexed Fluorescence Monitoring

Abstract:

Adsorption of polymers on single-walled carbon nanotubes (SWCNTs) has enabled developments in molecular sensing, *in vivo* imaging, and gene delivery applications.[1,2] Noncovalent functionalization of SWCNTs with polymers preserves the pristine SWCNT atomic structure, thus retaining the intrinsic near-infrared fluorescence for sensing or imaging functions, and offers a reversible binding mode for cargo delivery. However, noncovalent biomolecule adsorption is an inherently dynamic process, where exchange occurs between molecules in the bulk solution and molecules on the nanoparticle surface, into what is known as the corona phase. The nature, strength, and kinetics of both the polymer binding and unbinding processes on SWCNTs are important contributors to the success of polymer-SWCNT based technologies. Understanding this binding process is especially important for intended uses of functionalized SWCNTs in biological environments, where native biomolecules compete with the original polymer to occupy the SWCNT surface. Binding of proteins and other biomolecules to the SWCNT disrupts the intended functionality of the nanoparticle and can further lead to adverse biocompatibility outcomes.

We present an assay to study the corona exchange dynamics between solution-phase and corona-phase polymers on SWCNTs.[3] This assay exploits the quenching property of fluorophores proximal to the SWCNT surface to monitor ligand binding and unbinding events. Real-time tracking of polymer adsorption and desorption events on the SWCNT surface are conducted with systematic variation of the molecular entities and solution conditions, with a specific focus on DNA (as the sensing moiety) and protein (as the biofouling agent). Binding profiles are extracted from the experimental assay and used to inform a kinetic model of the system. The work presented herein develops an understanding of the fundamental corona exchange mechanism and provides insight into performance of these designed SWCNT-based systems in biologically relevant, protein-rich conditions.

References

1. Demirer, G. S. *et al.* High aspect ratio nanomaterials enable delivery of functional genetic material without DNA integration in mature plants. *Nature Nanotechnology* 1 (2019) doi:10.1038/s41565-019-0382-5.
2. Beyene, A. G. *et al.* Imaging striatal dopamine release using a nongenetically encoded near infrared fluorescent catecholamine nanosensor. *Science Advances* **5**, eaaw3108 (2019).
3. Pinals, R. L., Yang, D., Lui, A., Cao, W. & Landry, M. P. Corona Exchange Dynamics on Carbon Nanotubes by Multiplexed Fluorescence Monitoring. *J. Am. Chem. Soc.* (2019) doi:10.1021/jacs.9b09617.

598813 DNA-Caged Polymer Nanocomposites for Erasable Fluorescence Imaging

Abstract:

Imaging pathology samples is often difficult because of limited colorimetric dye options and tissue autofluorescence resulting from structural complexity. If fluorescent labels are used, they are limited to color channels with minimal overlap with tissue autofluorescence, which is not ideal for image multiplexing. Sequential erasable labeling with fluorescent antibodies has been proposed as a potential solution, often with the use of harsh chemicals to remove the antibodies. These chemicals can damage tissue, which may lead to loss of structural information or misidentification of important structures. Newer processes featuring single stranded DNA (ssDNA) modified with fluorophores as reporters may address these limitations. DNA labels can be easily erased by strand displacement reactions. However, unprotected fluorophores are easily bleached and lack stability. Here, we report the creation of DNA caged-micelles (i.e., DNA cages) for use as an erasable fluorescent labeling system. DNA cages are DNA nanostructures (modified from Kurokawa et al.[1]) that form a network on spherical surfaces (i.e., micelles). The proposed labeling system is based on gentle DNA dehybridization interactions and labels are protected by micelle coatings, which would greatly improve current technology. Additionally, the dyes used in cell and tissue labeling can include colorimetric, fluorescent, and quantum dot labels.

Proof of cage formation was measured using DNA tiles modified with fluorescent dyes and/or fluorescence quencher FRET pairs, as well as via TEM images. Three different fluorescent curves were generated based on interaction of tiles with the micelle surface and/or each other. Absorption of quenching DNA tiles to a fluorescent micelle surface indicated binding of DNA cages to the micelle surface, with complete quenching at a molar ratio of 13.5:1 (DNA:PEG (micelle)). This result was validated by TEM images showing formation of a new structure not observed in the absence of DNA or polymer (micelles). Additionally, cage saturation on the micelle surface was observed at a molar ratio of 18:1 and interlocking of the DNA tiles (based on a FRET quencher pair) was observed at a molar ratio of 9:1.

Erasable labeling was demonstrated in solution by binding the targeting strand on DNA cages to ssDNA labeled with fluorophores that was then erased in sequential cycles via strand displacement reactions. In solution, an erase depth of greater than 90% was achieved for multiple cycles. DNA cages were then loaded with fluorescent dyes and used as labels in fixed cells. U87 glioblastoma cells were labeled with primary antibodies targeting either integrin β1 (an extracellular target) or actin (an intracellular target). Then, secondary antibodies conjugated to ssDNA were used to bind primary antibodies. Finally, cages containing complementary ssDNA to that on the secondary antibodies were introduced for target labeling. Labeling of both targets was observed, consistent to that of gold standard controls. Next, DNA with greater complementarity to secondary antibodies was introduced, erasing DNA cages. An erase depth of up to 85% was achieved. DNA cages could also penetrate up to 100 μm in tissue sections with no damage observed over the course of treatment and washing. These results suggest that DNA cages show great promise for use in multiplexed cell labeling.

Reference

1 Kurokawa, C. *et al.* DNA cytoskeleton for stabilizing artificial cells. *Proc Natl Acad Sci U S A* **114**, 7228-7233, doi:10.1073/pnas.1702208114 (2017).

605587 Laser Induced Graphene on Polysulfone Membrane for H_2/CO_2 Gas Separation

Abstract:

There has been a growing interest in polymeric membranes for gas separation due to their low cost, low energy usage, and lack of moving parts. Dense polymeric membranes have an intrinsic trade-off between selectivity and permeability based on the kinetic diameter of the gases involved. Molecular sieving materials like graphene-based nanomaterials, MXenes, zeolites, and carbon nanotubes. shown promising applications in gas separation membranes. Mixed matrix membranes with improved selectivity over polymeric membranes have been developed by adding nanomaterials as fillers in polymer matrix. However, the membrane fabrication is limited by dispersion of fillers because of their incompatibility with polymer matrix and the membrane performance is compromised due to unselective voids. [1] Laser induced graphene (LIG) can be photothermally formed on surface of polymeric substrates; this method is fast and scalable and has been applied for ultrafiltration applications in polysulfone, polyethersulfone, and polyphenylsulfone.[2]

Here, we report that LIG on surface of polysulfone substrates can enhance the H_2/CO_2 perm-selectivity of the control polysulfone film. A new class of heterogeneous mixed matrix membranes is developed by forming graphene on surface of polymer and combining molecular sieving mechanism with solution-diffusion mechanism of baseline polymers. In a proof-of-concept experiment, we have used LIG on 100 micrometers thick polysulfone substrate for separation of light gas pair H_2/CO_2 at room temperature and 1 atm pressure. Preliminary measurements indicate that this combination is a high perm-selectivity membrane with a selectivity of 51 and permeance of 707 GPU (1 GPU = 3.35 x 10^{-10} mol/m^2/Pa) as compared to neat polysulfone with selectivity of 2 and permeance of 1 GPU. X-Ray Diffraction analysis, thermogravimetric data, optical and scanning electron microscopy were utilized to study the final material; these characterizations confirmed graphene formation on the substrate.

References:

[1] M. Etxeberria-Benavides, O. David, T. Johnson, M. M. Łozińska, A. Orsi, P. A. Wright, S. Mastel, R. Hillenbrand, F. Kapteijn, J. Gascon, *Journal of Membrane Science* **2018**, 550, 198.

[2] S. P. Singh, Y. Li, J. Zhang, J. M. Tour, C. J. Arnusch, *ACS Nano* **2018**, 12, 289.

610289 Study of Unique Thermodynamic Equilibria of Co-Surfactant Systems for Functionalizing Specific Chiralities of Single Walled Carbon Nanotubes

Abstract:

Single walled carbon nanotubes (SWCNTs) find applications in electronics and biosensors owing to their unique optical properties. Current biomedical applications are restricted due to constraints in separating highly monochiral fractions. Surfactant structure around the nanotube plays an important role in stability and separation of (n,m) species. The presence of thermodynamically stable co-surfactants states enables selective coating of species in solution. Our group has previously utilized this approach to elute (6,5) species at at very specific critical exchange ratio concentrations (CERC). The co-surfactant solution utilizes a ratio of sodium dodecyl sulphate (SDS)/sodium deoxycholate (DOC) to obtain single-chiral eluents of (6,5) species. This idea was further extended to obtain CERC values of different (n,m) species for the SDS-DOC co-surfactant system.

Here, we present utilization of this thermodynamic equilibria for selectively eluting a mixture of chiralities and modifying their functionality. Adsorption of added surfactant leads to reorganization of the surfactant shell around nanotube which leads to solvatochromatic shifts that can be probed through optical spectroscopy. Addition of DOC to SWCNT solution results in elution of a mixture of chiralities whose CERC is below the SDS/DOC ratio. Addition of SDS to this mixture results in opposite solvatochromatic shifts whose extent would indicate the surfactant reorganization around a particular (n,m) species. This would enable to functionalize selective species in solution. Different ligands were studied for functionalizing different chiralities for simultaneous detection of analytes in solution. This approach provides a promising pathway for using SWCNTs as biosensors by selectively functionalizing different chiralities without obtaining monochiral fractions.

598587 High-Yield Analysis of Individual Ions and Molecules through the Interior of Carbon Nanotubes

Abstract:

The interior channels of carbon nanotubes are promising for studying transport of individual molecules in a 1D confined space. However, experimental investigations of the interior transport have been limited by the extremely low yields of fabricated nanochannels and their characterization. Here, this challenge is addressed by assembling nanotube membranes on glass capillaries and employing a voltage-ramping protocol. Centimeter-long carbon nanotubes embedded in an epoxy matrix are sliced to hundreds of 10 μm-thick membranes containing essentially identical nanotubes. The membrane is attached to glass capillaries and dipped into analyte solution. Repeated ramping of the transmembrane voltage gradually increases ion conductance and activates the nanotube ion channels in 90% of the membranes; 33% of the activated membranes exhibit stochastic pore-blocking events caused by cation translocation through the interiors of the nanotubes. Since the membrane-capillary assembly can be handled independently of the analyte solution, fluidic exchange can be carried out simply by dipping the capillary into a solution of another analyte. This capability is demonstrated by sequentially measuring the threshold transmembrane voltages and ion mobilities for K^+, Na^+, and Li^+. Beyond the detection of solvated ions, varying the diameter of carbon nanotubes enables analysis of analytes with different sizes. We show, for instance, detection and identification of individual polymeric molecules with different chain lengths by using a carbon nanotube with diameter of ~3 nm. This approach, validated with carbon nanotubes, will save significant time and effort when preparing and testing a broad range of solid-state nanopores.

599430 Molecular-Dynamics Analysis of the Mechanical Behavior of Graphene Nanomeshes

Abstract:

Graphene nanomeshes (GNMs), also known as nanoporous graphene, are ordered defect-engineered graphene nanostructures consisting of periodic arrangements of nanopores in the graphene lattice with neck widths between pores of less than 10 nm, typically fabricated by block copolymer lithography and plasma/electron etching of graphene. The unique porous structure of GNMs and their outstanding mechanical and electronic properties makes these 2D materials very appealing for numerous technological applications, ranging from field effect transistors to membranes used in water desalination. Here, we present a systematic study of the response of GNMs to mechanical loading based on molecular-dynamics (MD) simulations according to a reliable bond order interatomic potential and establish the corresponding structure-properties relationships.

We report results on the response of graphene nanomeshes (GNMs) to uniaxial tensile straining based on MD simulations of dynamic deformation tests. We examine the effects on the GNM mechanical behavior under straining along different directions of the nanomesh pore morphology and pore edge passivation by testing GNMs with elliptical pores of various aspect ratios and different extents of edge passivation through termination with H atoms of under-coordinated edge C atoms. We establish the dependences of the ultimate tensile strength, fracture strain, and toughness of the GNMs on the nanomesh porosity, derive scaling laws for GNM strength-density relations, and find the GNMs' mechanical response to uniaxial straining to be anisotropic for pore morphologies deviating from circular. We also find that the GNM tensile strength decays exponentially with increasing GNM porosity and that pore edge termination with H atoms causes a reduction in the GNMs' elastic stiffening, strength, deformability, and toughness; this hydrogen embrittlement effect is more pronounced at a high level of pore edge passivation that renders the edge C atoms sp^3-hybridized.

We also report results on the mechanical and structural response of GNMs to indentation based on MD simulations of nanoindentation tests. We find the GNMs' response to indentation to be nonlinearly elastic until fracture initiation, with elastic properties that depend strongly on the GNM porosity but are not sensitive to pore edge passivation, which, however, influences the GNM failure mechanism past fracture initiation. Increasing GNM porosity leads to a monotonic decrease of the 2D elastic modulus of the GNMs and the modulus-porosity dependence follows a quadratic scaling law. The maximum stress reached at the GNM breaking point is high throughout the porosity range examined, even at very high porosity. The maximum deflection of the indented GNMs at their breaking point exhibits a minimum at porosities below 20%; beyond this critical porosity, the maximum deflection increases monotonically with increasing porosity and can reach values comparable to half of the indented sample radius at high porosities. Moreover, our analysis reveals a novel inelastic, dissipative necking mechanism of GNM failure at high porosities that further enhances the excellent deformability of the GNMs. Our findings highlight the potential of graphene nanomeshes as 2D mechanical metamaterials whose mechanical response can be tuned by proper tailoring of their structural features.

604900 Templating Colloidal Sieves for Tuning Nanotube Surface Interactions and Optical Sensor Responses

Abstract:

Surfactants offer a tunable approach for modulating the exposed surface area of a nanoparticle. They further present a scalable and cost-effective means for suspending single-walled carbon nanotubes (SWCNTs), which have demonstrated practical use as fluorescence sensors. Though surfactant suspensions show record quantum yields for SWCNTs in aqueous solutions, they lack the selectivity that is vital for optical sensing. We present a new method for controlling the selectivity of optical SWCNT sensors through colloidal templating of the exposed surface area. Colloidal nanotube sensors were obtained using various concentrations of sodium cholate, and their performances were compared to DNA-SWCNT optical sensors. Sensor responses were measured against a library of bioanalytes, including neurotransmitters, amino acids, and sugars. We report an intensity response towards dopamine and serotonin for all sodium cholate-suspended SWCNT concentrations. We further identify a selective, 14.1 nm and 10.3 nm wavelength red-shifting response to serotonin for SWCNTs suspended in 1.5 and 0.5 mM sodium cholate, respectively. Through controlled, adsorption-based tuning of the nanotube surface, this study demonstrates the applicability of sub-critical colloidal suspensions to achieve selectivities exceeding those previously reported for DNA-SWCNT sensors.

609319 Synergistic Effects of Carbohydrate Identity and Polymer Chain Length on Creating Glycopolymer-Wrapped Carbon Nanotubes

Abstract:

Dispersion of single-wall carbon nanotubes (SWCNTs) with biopolymers in aqueous environments facilitates their potential biological and biomedical applications. Glycopolymers, biomimetic polymers with carbohydrate pendant groups, have great potential for dispersion of SWCNTs since they have greater structural tunability and targeted biological functions compared to natural polysaccharides. In this work, we investigated a small library of synthetic glycopolymers with monosaccharide and disaccharide groups for dispersion of SWCNTs *via* non-covalent complexation in aqueous condition. We found that glycopolymers can effectively stabilize SWCNTs to form water-soluble glycopolymer-wrapped SWCNT complexes, depending on the carbohydrate density and polymer chain length. This work demonstrates for the first time the synergistic effects of the identity of carbohydrate pendant groups and polymer chain length of glycopolymers on stabilizing SWCNTs in water.

605158 Mechanistic Understanding of the Biological Responses to Polymeric Nanoparticles

Abstract:

Polymeric nanoparticles have demonstrated tremendous potential for theranostic and nanomedicine applications due their central role in the delivery of a wide range of therapeutic and imaging agents [1-4]. Nonetheless, increasing studies have reported the failures of these nanomaterials in realizing their designed objectives because of a single or combination of factors, notably non-specific biomolecule adsorption, inefficient cellular internalization, triggered immune response, and fast body excretion, which have collectively challenged the merit of nanomedicine for disease theranostics [5-7]. Although numerous efforts have attempted to regulate the biological behaviors of polymeric nanoparticles by controlling their physicochemical properties, a large number of the demonstrated studies have focused solely on either the biological effects induced through the regulation of a single nanoparticle parameter or the separate molecular, cellular, and *in vivo* behaviors of nanoparticles. Crucially, the interdependency of multiple physicochemical properties of polymeric nanoparticles in determining their biological effects from molecular to organism levels is still largely unclear, which warrants a thorough investigation.

In this work, we prepared a series of polymeric nanoparticles consisting of various biocompatible polymeric nanocarriers (i.e., polystyrene (PS), poly(lactic-*co*-glycolic acid) (PLGA), and polyethylene glycol (PEG)) encapsulating organic fluorophores [8]. Subsequent to the synthesis of these nanoparticles, we evaluated their physicochemical properties, such as surface morphology, size distribution, zeta potential, lipophilicity, optical absorbance, and fluorescence. Next, through a range of experimental characterization and molecular simulation, we systematically probed the interdependent effect of nanoparticle lipophilicity, zeta potential, and size on the formation of protein corona around the nanoparticles as well as the *in vitro* cellular uptake and *in vivo* biodistribution of nanoparticles in a zebrafish model.

We observed that nanoparticle lipophilicity influenced the recruitment of non-specific biomolecules, where the adsorption of biomolecules could be reduced by decreasing nanoparticle lipophilicity and vice versa. For a given lipophilicity, the nanoparticle size had a more significant effect in determining biomolecule adsorption than the nanoparticle surface charge. However, irrespective of size, the surface charge of nanoparticles was noted to affect their endothelium and macrophage uptake as well as circulation lifetime. More clearly, the negatively charged nanoparticles could be internalized preferentially by endothelial cells without the need for targeting ligands. Additionally, these nanoparticles were minimally internalized by macrophages and showed a much prolonged *in vivo* circulation lifetime. Based on our experimental data, we proposed a two-step framework to rationally design a single polymeric nanoparticle system with highly regulated *in vitro* and *in vivo* biological behaviors.

In summary, the optimization of the interplay between nanoparticle lipophilicity and surface charge is necessary to control the overall biological behaviors of polymeric nanoparticles. Our study has offered a strong basis for the engineering of enhanced polymeric nanoparticle systems to regulate multiple biological events simultaneously for *in vivo* theranostic and nanomedicine applications.

References

[1] Elsabahy et al., *Chemical Reviews* 115 (2015), 10967-11011.

[2] Ulbrich et al., *Chemical Reviews* 116 (2016), 5338-5431.

[3] Kenry et al., *Advanced Materials* 30 (2018), 1802394.

[4] Kenry et al., *Accounts of Chemical Research* 52 (2019), 3051-3063.

[5] Wilhelm et al., *Nature Review Materials* 1 (2016), 16014.

[6] Rosenblum et al., *Nature Communications* 9 (2018), 1410.

[7] *Nature Nanotechnology* 14 (2019), 1083.

[8] Kenry et al., *ACS Nano* 14 (2020), 4509-4522.

605108 Alterations in the Membrane Lipid Composition of Live Cells Affects the Ability of Cells to Internalize Nanomaterials

Abstract:

Over the past decades, engineered nanoparticles have been extensively used to deliver drugs or imaging agents to mammalian cells. Nanoparticles can be internalized into the cells through passive diffusion or a number of endocytic pathways. Previous studies on nanoparticle endocytosis have been primarily focused on elucidating the endocytic pathways, and the membrane receptors involved in each pathway. However, the role of the lipid microenvironment in regulating nanoparticle endocytosis has remained unexplored. In this study, live cell-lipid exchange was used to replace the lipids of the outer leaflet in A549, alveolar epithelial cells, with lipids of choice and examine their role in the uptake of silica nanoparticles (50 nm).

Methyl-alpha-cyclodextrin (MαCD) was used to alter the lipid composition of outer leaflet of cell plasma membrane. For the exchange studies, MαCD was pre-loaded with sphingomyelin (18:0, 18:1, and 24:0), 1,2-dioleoyl-sn-glycero-3-phosphocholine (36:2 PC, DOPC), and 2-((2,3-bis (oleoyloxy)propyl)dimethylammonio) ethyl hydrogen phosphate (36:2 inverse PC, DOCP). In each case, the exchange of cellular lipids with these synthetic lipids was confirmed by chromatography and mass spectrometry and cell viability after exchange was examined using the MTT assay. Control and lipid-exchanged cells were then exposed to fluorescent silica nanoparticle in serum-free medium or supplemented with fetal bovine serum (FBS) and incubated for one hour at 37 °C. The intensity of fluorescent nanoparticles, associated with or internalized into cell plasma membrane, was measured using flow cytometry. Lysosome and early endosome marker were used to study the intracellular fate of nanoparticles using confocal microscopy.

All exogenous lipids maintained cell viability; however, lipid-exchanged cells showed significant differences in their ability to endocytose nanoparticles in serum-free medium. DOCP and SM 18:0-exchanged cells showed significantly reduced nanoparticle uptake, SM 18:1-exchanged cells showed increased uptake, while DOPC-loaded cells showed a comparable level of uptake to control cells. Confocal microscopy confirmed the differences in observed in flow cytometry experiments, while showing differences in intracellular localization of nanoparticles in control and lipid-exchanged cells. No difference in nanoparticle uptake was observed in FBS-supplemented medium, likely due to the fact that serum proteins resulted in significant aggregation of nanoparticles, rendering them too large for endocytic uptake. These findings clearly indicate an important role for membrane lipid composition in nanoparticle uptake and are important in understanding the differences in uptake in cells with different membrane lipid compositions.

606677 Morphology-Based Transport of Gold Nanoparticles in Mature Plant Leaves

Abstract:

Plant genetic engineering has the potential to mitigate the effects of climate change and rising populations via the creation of high-yield crop varieties and climate-resilient plants. To this end, we require a plant genetic engineering toolset that is (i) plant-species independent, (ii) able to bypass physical barriers present in mature plant tissues, and (iii) capable of controlled localization and cargo release. A principal challenge in this field remains the efficient delivery of biological cargo across a plant's cell wall and cell membrane. Existing tools for plant genetic engineering suffer limitations in species range, cargo diversity, tissue damage, efficiency, and transgene integration resulting in regulatory oversight. Recently, nanoparticles (NPs) have emerged as promising materials for use as cargo carriers into plant cells. Owing to their highly tunable chemical and physical properties, NPs can be synthesized and surface functionalized to achieve targeted localization and cargo release.

Current techniques either use pathogens or biolistic force to deliver biomolecules to plants. There is interest in unassisted (non-biolistic, force-independent) delivery of genetic cargoes into plant cells, in which nanocarriers may need to remain below the cell wall size exclusion limit. However, there lacks a set of heuristics for NP design for unassisted DNA, RNA, and protein delivery in plants.

Herein, we utilize a library of NA-coated gold nanoparticles as model materials to elucidate the effect of NP morphology on transport in plant tissues and on cargo protection against nuclease degradation. Our study suggests that nanoparticle morphology impacts both the uptake pathway, timescale, and extent of nanoparticle internalization into plant cells. The workflow presented in this work can be utilized to probe the effect of other nanoparticle-based design parameters such as size and surface chemistry on biomolecule delivery efficacy, which can enable the creation of NP design heuristics for force-independent internalization. This work serves as an initial platform that can be built on to understand how design parameters affect delivery in plants for more targeted applications of bionanotechnologies in plant genetic engineering and more broadly in environmental and agricultural science.

596314 Silver Nanoparticle Surface Chemistry Affects Protein Interactions, Cell Viability, and Gene Expression in Human Liver Cells

Abstract:

Silver nanoparticles (AgNPs) are increasingly being used in consumer and biomedical applications for their antimicrobial and plasmonic properties. Yet, detailed characterizations of physiological responses to AgNPs, as well as the influence of particle properties (such as surface chemistry) and interactions with other biomolecules on these responses are needed before their wide application. Furthermore, increasing concerns of frequent and cumulative exposures point to a need for cumulative risk assessment studies of AgNP exposure. Our work uses AgNPs with a range of surface coatings to evaluate the role of surface charge in protein interactions, cellular cytotoxicity, and gene expression. Conformational changes in the model protein human serum albumin (HSA) upon AgNP interaction, as monitored via circular dichroism spectroscopy, were most pronounced for cationic particles. In contrast, HSA-AgNP binding constants were highest for negatively charged AgNPs, though the differences in affinities were not significant across different surface coatings. Consistent with the trends in HSA-AgNP binding affinities, HSA-mediated changes in AgNP hydrodynamic radius and oxidative dissolution were highest for anionic particles. Cytotoxicity experiments carried out using hepatic cells revealed that none of the AgNP preparations studied showed significant toxicity at low nanoparticle concentrations. At higher concentrations, however, cationic AgNPs showed significant impacts on cell viability, both with and without HSA coating. Consistent with the cytotoxicity studies, results of transcriptomics analysis demonstrated that AgNPs altered gene transcription of HepG2 cells in a time-dependent manner, with the most dramatic effects seen for cationic AgNPs. Furthermore, universal to all surface-coatings, AgNPs treated cells responded by inactivating proliferation and enabling cell cycle checkpoints. Taken together, these results demonstrated the significance of AgNP surface charge in both biophysical HSA-AgNP interactions and cytotoxicity and gene expression, with cationic AgNP causing greatest disruption to HSA structure and cell responses.

609405 Inert Nanoparticles for Enhancing the Survival of Primary Macrophages.

Abstract:

Cell-based therapies have recently emerged as a new platform for treating a spectrum of chronic degenerative diseases, with serious advances made especially in hematopoietic stem cell and reactive lymphocyte therapies. Macrophages are leukocytes responsible for phagocytosing foreign objects, bacteria, and apoptotic cells, maintaining homeostasis, and bridging the innate and adaptive immunity. Macrophage-based cell therapy has been identified as a promising therapeutic approach to treat chronic immune dysfunctions, owing to the functionality and phenotype plasticity of macrophages. However, a major challenge in the advancement of macrophage-based cell therapies is the difficulty to maintain adequate cell viability *ex vivo* following macrophage isolation from tissue. Current attempts of macrophage *in situ* activation or autologous cell therapy following *ex vivo* stimulation fail to elicit effective therapeutic responses, due to low cell survival rates upon reintroduction. Therefore, there is a need to develop new approaches to regulate primary macrophage survival in both *in vivo* and *ex vivo* environments.

We attempted to extend the viability of *ex vivo* macrophage cultures through treatment of inert poly(ethylene glycol) (PEG) diacrylate (PEGDA)-based nanoparticles. We observe that *ex vivo* bone marrow-derived macrophages (BMMs), which typically undergo apoptosis within 2 weeks under appropriate culture conditions, can persist and maintain functionality for up to 6 months after being treated with a *single* dose of PEGDA nanoparticles. We find that nanoparticle uptake significantly delays primary macrophage cell death through the downregulation of caspase-dependent apoptosis pathways. Furthermore, we observe no major changes to macrophage phenotype following nanoparticle uptake. However, phagocytosis of PEGDA nanoparticles leads to increased expression of major histocompatibility complex (MHC) class II uniquely in the absence of other co-stimulatory molecules. The enhanced survival phenomenon is also applicable to *ex vivo* cultures of terminally differentiated macrophages isolated from tissue, including peritoneal and alveolar primary cells. Overall, this work demonstrates for the first time the ability of inert nanoparticles to suppress apoptotic signaling and prolong viability in primary macrophages extracted from different tissues. This work could eliminate a major obstacle standing in the way of developing macrophage-based cell therapies, as well as provide mechanistic insight into the implications of nanoparticle phagocytosis on cell longevity and function.

612225 Peptides As Surface Coatings of Nanoparticles That Penetrate Human Cystic Fibrosis Sputum and Uniformly Distribute In Vivo Following Pulmonary Delivery

Abstract:

In mucosal-associated diseases, drug and gene delivery systems have to traverse multiple biological barriers to achieve therapeutic efficacy. In particular, in select pulmonary diseases such as cystic fibrosis (CF), therapeutic delivery remains a significant challenge due to concentrated viscoelastic mucus, which prevents effective penetration and transmucosal transport of drugs and particles. To address this problem, we used combinatorial peptide-presenting phage libraries with next-generation DNA sequencing to identify hydrophilic, net-neutral charged peptides as coatings that enable penetration of human CF mucus *ex vivo* with ~600-fold better penetration than control. In addition, these peptide coatings improve uptake into lung epithelial cells compared to uncoated or standard poly(ethylene glycol) (PEG)-functionalized nanoparticles and exhibit enhanced uniform distribution and retention in the mouse lung airways. These peptide coatings address multiple delivery barriers and effectively serve as excellent alternatives to standard PEG surface chemistries to achieve mucus penetration and address some of the challenges encountered using these chemistries. This biomolecule-based strategy can address multiple delivery barriers and hold promise to advance efficacy of therapeutics for diseases like CF.

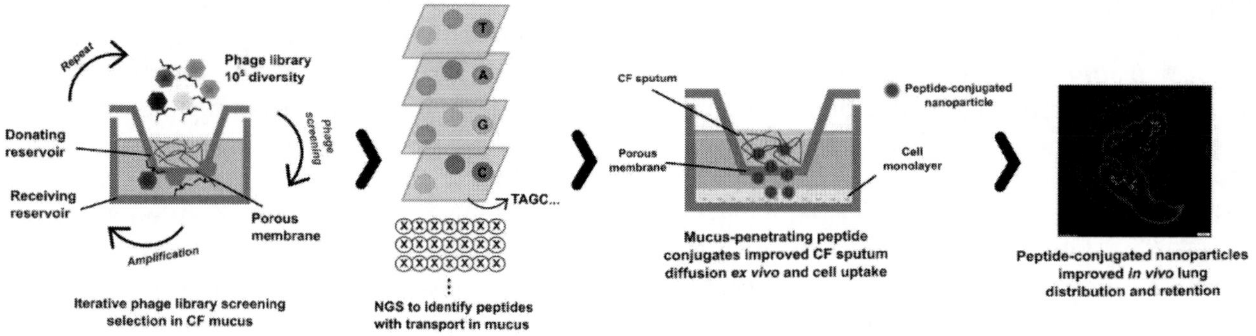

603892 (Invited) Skin-like Wearable and Implantable Sensors

Abstract:

In this talk, I will present our recent work on BodyNet concept for stretchable wearable and implantable sensors. Examples include, full body sensing, blood flow sensor and morphing electronics - neurostimulator.

600113 An Adhesion-Based Anti-Corrosion Strategy for Wearable Electrochemical Sensing and System Integration

Abstract:

The large-scale deployment of biomonitoring sensors through integration in wearable consumer electronics enables population-level health and wellness monitoring, and thus can play a critical role in transforming personalized medicine. Currently, physical sensors have been widely incorporated within commercialized wearable platforms, but they cannot capture biomarker information that lie in molecular levels and provide highly specific information about the body's dynamic chemistry. Electrochemical biosensing interfaces can be positioned to target biomarker molecules in a wearable format, as they can be constructed on miniaturized/flexible footprints/substrates and can convert (bio)chemical signals into electrical signals measurable by electronic readouts. The contact pads on the printed circuit board (PCB), which is a core component of wearable consumer electronics, can directly serve as electrodes and as a substrate for electrochemical sensors. However, we found that the commercialized PCB metal contact pads (*e.g.*, gold) were prone to corrosion and unsuitable for electrochemical modifications in solution environments. Additionally, electrochemical biosensors often are not reusable and need to be replaced frequently, while PCB units are not positioned for disposable applications.

To this end, we devise an adhesion-based anti-corrosion strategy for PCB-interfaced electrochemical sensing, which enables seamless sensor-system integration. The strategy is based on decoupling the sensing interface and PCB metal contact pads by an intermediary ultrathin (~50 μm), vertical conductive, and double-sided adhesive film. This strategy has dual benefits of making electrochemical sensors disposable and insulating the contact pads from sounding solutions. We first establish the suitability of the adhesion film to serve as an electrode and as an interconnection film simultaneously. One side of the double-sided adhesive film was patterned with noble metal electrodes, followed by the modifications with recognition layers, such as permselective membrane and enzymes, to render biosensing functionality. The electrochemical biosensors were transferred and adhered to the PCB contact pads with the aid of water-soluble tape. As compared with bare PCB contact pads, this fabrication strategy allows for realizing a corrosion-resistant and stable electrochemical biosensing interface on PCB for operation in biofluid (*e.g.*, sweat). Amperometry and open circuit potential characterization experiments validated the electrochemical corrosion resistance properties in saline solution environments. The excellent anti-corrosion attributes of the modified film could be ascribed to: (1) the polymer glues on both sides of the adhesive film acting as corrosion-protective layers; (2) bypassing the need for oxidation prone-adhesion metal (*e.g.*, chromium), achieved by e-beam evaporation (at the noble metal patterning step); and (3) the electrodeposited permselective membrane (here, phenylenediamine), which has been reported to presented anti-corrosion properties. We specifically tailored the enzymatic layer of the electrochemical interface to target biomarker molecules such as glucose, lactate, and choline as example of informative metabolites and nutrients. By modifying different (bio)chemical layers various other molecules can be targeted. In this way, our strategy serves as a foundation for electrochemical sensor development and integration with wearable consumer electronics.

607021 Rapid Screening Platform for the Isotopic Determination of Radionuclides in Water

Abstract:

Traditional approaches for the detection of trace-level radioisotopes in water require lengthy offsite sample preparations and do not lend themselves to rapid screening. Therefore, a novel sensor platform that combines onsite purification, concentration, and isotopic screening with a fieldable detection system will be an invaluable tool.

In this presentation, we will discuss findings on the development of a sensor platform based on reactive polymer film coatings for the rapid isotopic screening of waterborne radionuclides. Nanothin coatings were prepared by grafting uranium-selective polymers from polyethersulfone (PES) surfaces via UV-initiated polymerization, and by introducing uranium-selective functional groups to polyacrylonitrile (PAN) surfaces by chemical reaction. Ellipsometry was used to study film growth kinetics on PES films. X-ray photoelectron spectroscopy of modified PAN films revealed conversion of nitrile groups to amidoxime groups to be as high as 40% and showed that the extent and depth of reaction could be varied precisely. Static uptake experiments with solutions of depleted uranium spiked with ^{233}U were conducted to determine uranium binding capacities and kinetics of the modified polymer films at different pH values from 4 to 8. Sorption isotherm data were fitted to the Langmuir model, and the highest sorption capacities were obtained at pH 6 for modified PAN (M-PAN) and PES (M-PES) films. Capacities at pH 4 and 8 were lower and could be explained by differences in sorption mechanisms. Uranium batch uptake kinetics followed a pseudo-second order rate model. Alpha spectroscopy pulse height spectra were analyzed to study the role of selective layer film thickness on peak energy resolution. Excellent resolution with full width at half maximum values from 29 to 41 keV were recorded for M-PAN film and from 26 to 45 keV for M-PES film. Whereas uranium uptake increased with selective layer film thickness and varied with polymer chemistry/extent of modification, the peak energy resolution was independent of layer thickness and polymer chemistry within the experimental measurement uncertainties.

Results from this work were used to modify PES and PAN ultrafiltration membranes with the same experimental conditions to develop uranium selective membranes. High-throughput uptake experiments were performed to study the effect of residence time on detection efficiency and isotopic resolution. This membrane-based detection method enables the rapid, fieldable screening analysis of radionuclides in water for environmental monitoring studies and nuclear forensics investigations.

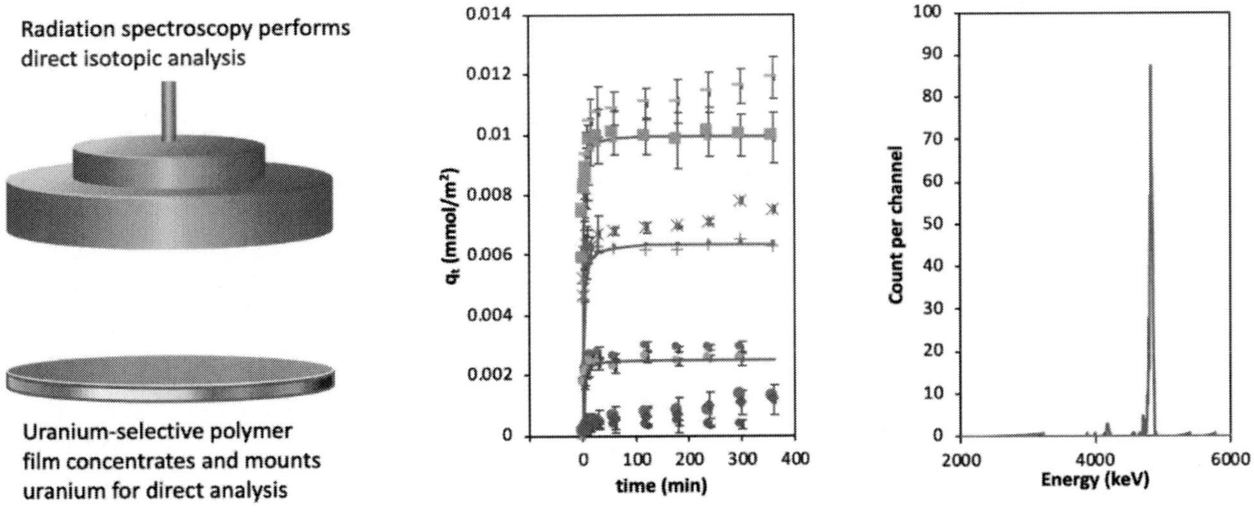

600875 Electrochemical Detection of Environmental Pollutants

Abstract:

Many small molecule pollutants interact with human hormonal systems. These compounds, known as endocrine disruptors, are ubiquitous, as they are found in plastics, medications, and pesticides. Such chemicals are especially problematic because they have been linked to severe health problems, including cancer, infertility, and diabetes. We have developed two complementary electrochemical methods to rapidly detect and quantify these compounds. One technique involves the specific monitoring of Bisphenol A (BPA) using a DNA aptamer immobilized on an electrode. The second method involves the measurement of the biological activity of chemically dissimilar endocrine disruptors using a human hormone receptor as a component of an electrochemical sandwich assay. These electrochemical platforms require no specialized skills to implement and have enabled very sensitive detection (sub-ppb levels) of these environmental pollutants from complex solutions, such as baby formula.

604144 Fabrication of High-Resolution Graphene-Based Flexible Electronics Via Polymer Casting and Microfluidic Approaches

Abstract:

Graphene-based flexible electronic devices for sensor applications has been receiving growing interest due to unique properties of graphene.[1] Various fabrication methods, such as photolithography, printing and chemical vapor deposition (CVD), has been used to produce high resolution and low feature size graphene-based devices.[2, 3] However, most of these methods are complex and require multiple processing steps (i.e. stamping, vacuum drying and etching) as well as additional post treatments (i.e. high temperature (300-1000 °C) baking or laser annealing) that can thermally or chemically degrade polymer-based substrates limiting substrate material selection.[2, 3] In addition, these methods can only be applied to 2D planar substrates and are not feasible to fabricate 3D circuits and some of them, such as sticky/adhesive tape peeling method, require high amounts of graphene consumption.[4] Despite the recent progress in the field, there is still a need to develop novel fabrication approaches that can address the current limitations of the existing methods.

In this study, we present two flexible electronics device fabrication methods based on simple polymer casting and microfluidics approaches.[5, 6] The first method focuses on simple polymer casting based graphene transfer at room temperature that does not require any post-processing. Briefly, our method consists of two main steps; (i) the formation of graphene patterns/films on substrates/molds via conventional methods such as CVD, channel filling or ink-jet printing and (ii) direct casting of target substrate polymer solution on the substrates/molds with graphene micropatterns and direct graphene transfer to the target substrate via peeling off upon drying and film formation. This method simply relies on the differences in the surface energies and adhesive forces between the graphene/mold and graphene/target polymer substrate.[5] Alternatively, the second method concentrates on a novel microfluidic approach to fabricate high resolution and low feature size graphene-based circuits/devices.[6] This method involves controlled and selective filling of microchannels on substrate surfaces with a conductive binder-free graphene nanoplatelets (GNP) solution. The ethanol-thermal reaction of GNP solution at low temperatures (~75 °C) prior to microchannel filling (as called "pre-heating") potentially reduces and homogenize GNP, which in turn enhances conductivity and reduces sheet resistance (down to ~0.05 kΩ sq^{-1}). Therefore, harsh post-processing of GNP is eliminated, and room temperature fabrication is enabled, which allows the use of degradable substrates, including biodegradable and natural polymers. This simple process is followed by selective administration of conductive GNP solution to the predetermined microfluidic channels on flexible substrates in a controlled manner via a syringe pump to create high resolution and low feature size graphene circuits/devices. This microfluidic approach can also be used to create 3D circuits with different geometries, using combined approaches with 3D printing, which is difficult to obtain with other methods (i.e. inkjet, screen or aerosol printing) that work mainly with 2D planar substrates.[6]

We demonstrated that using these fabrication methods, we obtained high resolution and low feature size graphene circuits with different geometries, dimensions and 3D superficial microstructures. We investigated the effect of temperature and microfluidic parameters (such as concentration, volume and flow rate) as well as the type of substrate materials on the graphene structure and formed micropatterns. The XPS, Raman, SEM and TEM results validated the structural changes of graphene along with confirming the stable presence of graphene patterns on the target substrates.[5, 6] The feature sizes of the graphene patterns could range from a few micrometers (down to ~15 μm in width and ~5 μm in depth) to a few millimeters with very small amounts of GNP used (~2.5 mg of graphene to obtain ~0.1 kΩ sq^{-1} of sheet resistance for 1 cm^2). [5, 6] The generated graphene patterns demonstrated significant stability after multiple washing and bending cycles. These methods were also implemented using other conductive inks, such as conductive graphene-silver based composite ink.[5, 6] Room temperature processing also provided precise control of 3D microstructural and mechanical properties (such as film porosity, pore size, elasticity etc.) of the target substrate materials, as demonstrated by SEM images.[5, 6] Using these methods, we successfully fabricated pressure sensors and near field communication antennas. [5, 6] Our results demonstrated that the devices worked with high sensitivity and stability.

In conclusion, the circuits/devices fabricated using this method can easily be implemented as wearable electronic, sensors and batteries, robotic components or motion detectors. In a broader sense, these simple, environmentally friendly and low-cost methods could pave the way for the fabrication of 2D and 3D electronic circuits/devices on different degradable flexible substrates using various graphene-based conductive inks.

References:

1. Jang, H.; Park, Y. J.; Chen, X.; Das, T.; Kim, M. S.; Ahn, J. H., Graphene-Based Flexible and Stretchable Electronics. *Advanced Materials* **2016,** *28* (22), 4184-4202.
2. Das, S. R.; Nian, Q.; Cargill, A. A.; Hondred, J. A.; Ding, S.; Saei, M.; Cheng, G. J.; Claussen, J. C., 3D nanostructured inkjet printed graphene via UV-pulsed laser irradiation enables paper-based electronics and electrochemical devices. *Nanoscale* **2016,** *8* (35), 15870-15879.
3. Das, S. R.; Srinivasan, S.; Stromberg, L. R.; He, Q.; Garland, N.; Straszheim, W. E.; Ajayan, P. M.; Balasubramanian, G.; Claussen, J. C., Superhydrophobic inkjet printed flexible graphene circuits via direct-pulsed laser writing. *Nanoscale* **2017,** *9* (48), 19058-19065.
4. Oren, S.; Ceylan, H.; Schnable Patrick, S.; Dong, L., High-Resolution Patterning and Transferring of Graphene-Based Nanomaterials onto Tape toward Roll-to-Roll Production of Tape-Based Wearable Sensors. *Advanced Materials Technologies* **2017,** *2* (12), 1700223.
5. Uz, M.; Jackson, K.; Donta, M. S.; Jung, J.; Lentner, M. T.; Hondred, J. A.; Claussen, J. C.; Mallapragada, S. K., Fabrication of High-resolution Graphene-based Flexible Electronics via Polymer Casting. *Scientific Reports* **2019,** *9* (1), 10595.
6. Uz, M.; Lentner, M. T.; Jackson, K.; Donta, M. S.; Jung, J.; Hondred, J.; Mach, E.; Claussen, J.; Mallapragada, S. K., Fabrication of Two-Dimensional and Three-Dimensional High-Resolution Binder-Free Graphene Circuits Using a Microfluidic Approach for Sensor Applications. *ACS Applied Materials & Interfaces* **2020,** *12* (11), 13529-13539.

607805 Synthesis and Characterization of Nano-Ettringite for Nano-Engineering of Cementitious Composites

Abstract:

The use of nanomaterials in modifying cementitious binders has gained tremendous attention in recent decades. Various nano-sized anhydrous inorganics, e.g. nano-silica, nano-titania, nano-limestone, and nano-clay etc., have been studied and shown to impact positive effects on both the fresh and mature state properties of cement composites. To the best of our knowledge, no studies have been conducted to investigate the use of nano-sized cement hydration products (e.g., nano-ettringite) in Portland cement or calcium sulfoaluminate binders. Ettringite is one of the first phases produced during the hydration of Portland cement, and it is the principal phase in hydrated calcium sulfoaluminate cements. It plays a vital role in controlling the early-age property of cementitious systems. Controlling the rate of ettringite precipitation can allow builders to regulate the hydration and physico-mechanical properties of cementitious binders according to their requirements.

This study aims at the synthesis of nano-sized ettringite for use in nano-engineering of concrete binders. A chemical precipitation method was studied with focus to understand the effects of various process parameters like addition of surfactants, stirring speed, and reaction time on the nucleation and growth of ettringite nanocrystals. Synthesized ettringites were characterized by dynamic light scattering (DLS), powder X-ray diffraction (XRD), scanning electron microscopy (SEM) and infrared spectroscopy techniques. Results show that ettringite nano-rods readily nucleate and start growing within the first 30 minutes of mixing stoichiometry proportions of calcium hydroxide and aluminum sulfate precursor solutions. Although, the addition of surfactant and stirring parameters were observed to influence the size of synthesized ettringite nanoparticles, the reaction time clearly plays a dominant role in determining the size of the synthesized ettringite.

608136 Flow Reactors for Rapid Screening of Reaction Parameters to Synthesize Molybdenum Carbide Catalysts for Biomass Conversion Processes

Abstract:

Catalytic conversion of biomass has garnered significant research attention to promote the development of sustainable routes to fuels and chemicals. This led to more effort toward finding applicable, earth-abundant, and cost-effective catalysts that promote selective transformation of biomass-derived compounds into useful products. Nanostructuring of transition metal carbides is an outstanding candidate for multiple catalytic transformations (e.g. hydrogenation, deoxygenation, C-C coupling) of biomass-derived compounds due to their superior catalytic performance. Recently, solution-phase phase pure α-MoC_{1-x} catalysts were manufactured by thermolytic decomposition of $Mo(CO)_6$ at mild reaction conditions for thermocatalytic CO_2 hydrogenation. However, syntheses in batch processes hinder the control and optimized tuning of the nanoparticles' physical properties, which may affect catalytic performance. In order to investigate the reaction parameters and their influence on the resulting nanoparticles, we have implemented a millifluidic flow reactor system for screening the experimental variable space for MoC_{1-x} nanoparticle synthesis across a wide range of reaction conditions. The superior heat and mass transport characteristics provided by millifluidic flow reactors allows rapid parametric screening and continuous monitoring of the experimental variables to tune nanoparticle properties. Design of Experiment analysis is preformed to assess the influence of reactor temperature, flow rate, and ligand concentration to find the parameters that have statistically significant effects on the response variables, namely, the residence time, yield, and nanoparticle size and morphology. Performing parametric screening for controlled nanostructuring including size, morphology, and crystallinity may lead to greater understanding of the structure-function relationships of the resulting nanoparticles and provide an optimal tool to tune nanoparticle properties.

608865 Carbon — I-Polypropylene Nanocomposites Synthesized with Zirconocene/MAO Catalyst System: Property Estimation through Simulation of De Assisted Dendritic Neuron Model

Abstract:

ABSTRACT

Industrial and academic research has shown a huge interest in polyolefin nanocomposites in recent years. These nanocomposites are high potential materials with much improved mechanical properties sch as strength, toughness, stiffness and flame retardancy.

Single site catalysts, with the common name metallocenes, are increasingly being studied and used to produce various terminal-olefinic polymers with more specific molecular architecture than those can be synthesized by conventional polymerization methods like using Zeigler-Natta type catalysts. Metallocenes, being single site catalysts, permit the synthesis of polymer nanocomposites with tailored microstructural properties and control over tactility and stereoregularity. Generally, the nano-dimensional component makes the reinforcement phase in polymer nanocomposites. These reinforced nanocomposites show a radical enhancement to the physical and chemical properties to the native polymer matrix.

The use of dendritic neural networks (NNs) has become gradually prevalent for applications where the automatous portrayal of the interrelation of dependent and independent variables is either incomprehensible or very complex.[1] The neural network consists of processing neurons and information flow channels between the neurons, usually called 'interconnects'. Each processing neuron calculates the weighted sum of all interconnected signals from the previous layer plus a bias term and then generates an output through its activation transfer function. Training is done by assigning random weights to each neuron, evaluating the output of the network and calculating the error between the output of the network and the known results by means of an error or objective function. If the error is large, the weights are adjusted, and the process goes back to evaluate the output of the network. This cycle is repeated till the error is small or a stop criterion is satisfied.[2] While training an artificial NN, one needs to find out optimal weight and bias set of the network. Currently, many traditional optimization techniques, such as Backpropagation algorithm, Quasi-Newton algorithm, Genetic algorithm, Simulated Annealing algorithm, Particle Swarm Optimization algorithm *etc.* are widely used to train the neural networks. Differential evolution (DE), is one among the non-traditional optimization technique and it has proven, through benchmark tests, to be superior to many algorithms in terms of speed and convergence.[3]

DE is a population based, stochastic, evolutionary algorithm, capable of finding global optima for complex optimization problems that are nondifferentiable, non-linear, discrete or multi-modal in nature, with only a few parameters.

This work deals with differential evolution assisted artificial dendritic neuron modeling for property estimation of carbon nanocomposites of iso-polypropylene synthesized with $Me_2Si(2—Me-Ind_2)ZrCl_2/MAO$ catalyst system. The effect of temperature, monomer concentration and catalyst concentration with different fillers on catalyst activity, molecular weights and melting point of the carbon nanocomposites of iso-polypropylene synthesized with zirconocene catalyst has been investigated in MATLAB® (R2020a) software.[4]

References

[1] Fernandes F. A. N. and Lona L. M.F., *Brazilian Journal of Chemical Engineering*, 22: 03, p. 401-418 (2005).

[2] Haykin, S., *Neural Networks: A Comprehensive Foundation*, Prentice Hall, New York (1998).

[3] Vesterstrom, J. & Thomsen, René., *Congress on Evolutionary Computation*, 980-987 (2004).

[4] MATLAB. version (R2020a). Natick, Massachusetts: The MathWorks Inc. (2020).

598171 Mechanism of Nanoparticle Synthesis in Reversible Ionic Liquid Systems

Abstract:

Nanoparticle synthesis with silylated amine reversible ionic liquids (RevILs) has been previously demonstrated to offer a unique alternative to traditional synthetic methods allowing for size control and facile deposition on to support surfaces; however, the mechanism of nanoparticle synthesis in these systems remains uncharacterized. Previous work hypothesized nanoparticles prepared with RevILs are formed via a reverse micelle mechanism in which nanoparticles are stabilized and templated within the aqueous core of the organized structures. In this work, diffusion-ordered nuclear magnetic resonance spectroscopy (DOSY-NMR) is used to demonstrate that nanoparticles synthesized with 3-aminopropyltriethylsilane (APTES) RevIL are not formed through a reverse micelle mechanism but rather a switchable aggregation of the silylated amine that affords control over nanoparticle size via manipulation of RevIL structure and concentration. Diffusion measurements and ^{1}H NMR spectra indicate that there is no aqueous core present within RevIL structures; however, nanoparticle synthesis occurs within the RevIL aggregates allowing for control over nanoparticle size via manipulation of aggregation number. Further, it is shown that the addition of water to these systems has detrimental effects on the aggregation behavior of the ionic liquid molecules in solution causing disassembly of the ion pairs.

611050 Continuous Tannin-Mediated Synthesis of Silver Nanostructures

Abstract:

Green methods for the synthesis of silver nanostructures have been receiving significant attention due to their non-hazardous, environmentally-friendly, and low-cost characteristics. Silver nanostructures can be synthesized using bacteria, fungus, yeast, viruses, vitamins, algae, plants and other natural compound or microorganism containing functional groups with reduction and capping capabilities. However, unlike conventional chemical and physical methods, the challenges in green synthesis process are broad size distribution, non-homogeneous morphologies, and low yield of synthesized silver nanostructures. Synthesis using plant extracts has emerged as a promising versatile method for improvement of the synthesis processes, as an alternative for current chemical and physical methods. However, the mechanism through which the reduction, nucleation, and nanoparticle growth occur is still obscure.

In this work, by using tannin as a strong reducing agent, we aim to demonstrate a facile and low-cost method for the synthesis of silver nanostructures and describe what are the most significant factors controlling the synthesis process using both in-situ and ex-situ characterization techniques. We show, how the size and morphology of the synthesized silver nanostructures change based on different reaction conditions. The in-situ Fourier-transform infrared spectroscopy (FTIR) is used to observe how different functional groups are changing during the course of reaction. In addition to that, Transmission Electron Microscopy (TEM), Scanning Electron Microscopy (SEM), Energy Dispersive X-Ray (EDX), and Ultraviolet–visible Spectroscopy (UV-vis) are used to characterize size and morphology of the synthesized silver nanostructures. The findings from these characterization techniques are essential to find the reaction mechanism behind the synthesis process.

It is found that the process does not need any capping agent to direct the anisotropic growth of the nanoparticles to form one-dimensional (1D) silver nanostructures such as nanorods (NRs) and Nanowires (NWs). However, the 1D silver nanostructure yield is low, and we illustrate how the application of continuous millifluidic reactors can increase their yield and can overcome the challenges in batch synthesis through control of a uniform reaction environment in a small reaction volume. These novel reactor systems offer similar advantages as microfluidic ones while being easier to fabricate, simpler to reconstruct, and even more adaptable for in-situ monitoring characterizations. Larger surface to volume ratio and precisely controlled flow patterns that consequently increase heat and mass transfer rate, coupled with inherent safety are among other advantages of flow millifluidic platforms.

611902 Colloidal Nanoparticle Catalysis

Abstract:

Gold nanoparticles (AuNPs) have attracted enormous attention as of late due to their unique catalytic activities. Colloidal AuNPs provide benefits of selectivity, greater surface area per mass of catalyst compared to supported catalysts, activity under mild conditions and potential for chiral catalysis. We have demonstrated that colloidal AuNPs have higher catalytic activity than the supported counterparts on a per metal basis. The caveat is that colloidal AuNPs require surface ligand functionalization in order to prevent aggregation and loss of activity. This results in surface passivation and significant reduction in catalytic activity. Colloidal catalytic activity is strongly dependent on ligand packing and conformation on the AuNP surface. High molecular weight polymer ligands provide increased available surface area owing to decreased surface coverage due to steric hindrance. This leads to increased catalytic activity, while low molecular weight ligands lead to complete AuNP surface passivation. A major challenge for colloidal AuNP catalysts is the catalyst recovery and reuse. To alleviate this, we have developed pH responsive ligands that enable recursive recovery and reuse of colloidal AuNP catalysts by altering the pH to selectively precipitate or phase transfer catalyst into organic solvents. This was achieved by employing poly(acrylic acid) as benchmark polymer ligand where a specific number of acid groups can be grafted to with thiol moieties for surface binding. In this case there is a fine balance between providing adequate nanoparticle colloidal stability, recoverability, and maintaining catalytic activity.

611384 Plasmon-Enhanced Greenhouse Effect

Abstract:

The greenhouse effect occurs when a material transmits sunlight and restricts thermal radiative losses by forcing thermal photons to be absorbed and re-emitted many times before escaping. This mechanism explains why increasing concentrations of gases such as methane and carbon dioxide in the atmosphere lead to global warming. Concentrated solar thermal (CST) technologies can utilize the greenhouse effect for carbon-neutral electricity and heat generation, but its effect is too weak for efficient use. Here, we report a significant improvement in the greenhouse effect by leveraging local surface plasmon resonances within hollow three-dimensional structures. Specifically, we incorporate nanoparticles of a transparent conducting oxide into a silica aerogel matrix and characterize their impact on solar and thermal transport. We also develop a new definition for greenhouse selectivity – a quantitative measure of the greenhouse effect – and relate it to the performance of solar receivers for CST. The plasmon-enhanced greenhouse selectivity (PEGS), demonstrated here, translates into a nearly two-fold reduction in measured heat losses at high temperatures (>600 deg. C) and enables higher operating temperatures for future CST applications.

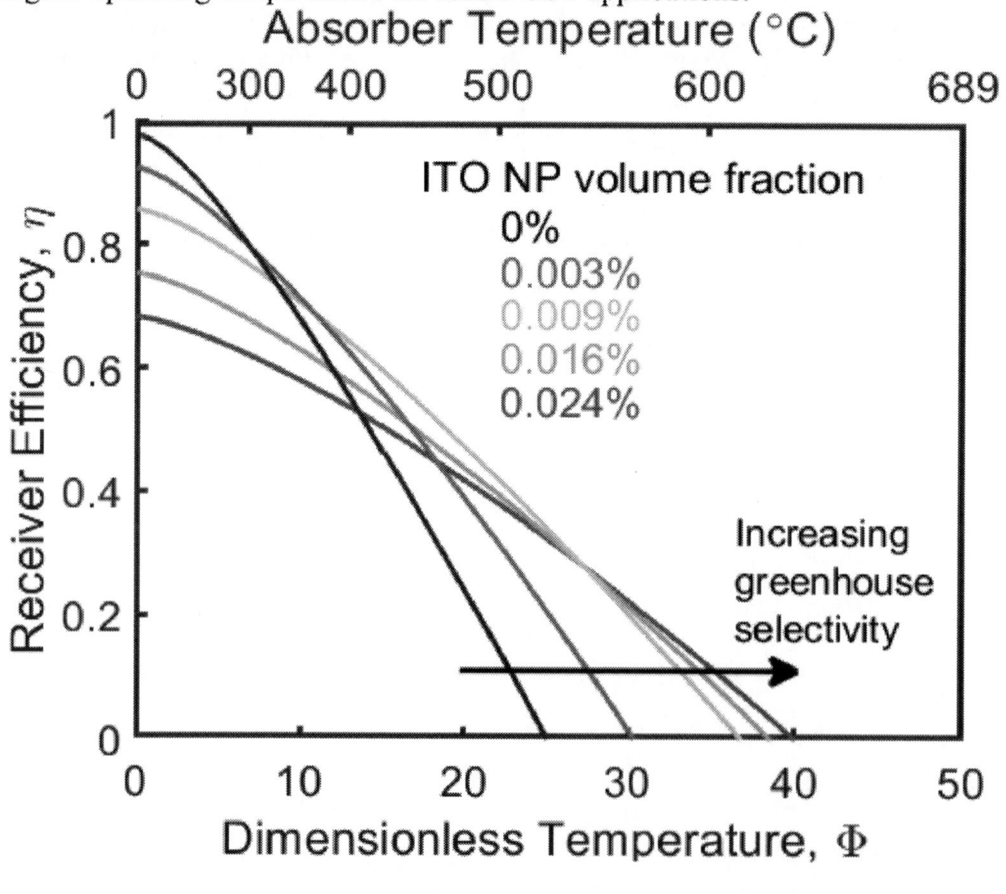

610815 Ultrafast Relaxation Dynamics in Bimetallic Plasmonic Catalysts

Abstract:

Combining a plasmonic metal, such as gold, with other popular catalysts, such as Ni or Pt, can extend its benefit to many energy-extensive reactions catalyzed by those metals. The efficiency of a plasmon-enhanced catalytic reaction is mainly determined by the light absorption cross section and the photoexcited charge carrier relaxation dynamics of the nanoparticles. We have investigated the charge carrier relaxation dynamics of gold/nickel (Au/Ni) and gold/platinum (Au/Pt) bimetallic nanoparticles. We found that the addition of Ni or Pt to gold can reduce light absorption in gold nanoparticles. However, electron-phonon coupling rates in Au/Ni and Au/Pt nanoparticles are significantly faster than that of pure Au nanoparticles. This is due to the fact that both Ni and Pt possess significantly larger electron-phonon coupling constants and higher densities of states near the Fermi level in comparison to Au. Additionally, the phonon-phonon coupling rate of bimetallic Au/Pt and Au/Ni nanoparticles was significantly different than that of pure gold nanoparticles, due to the acoustic impedance mismatch at the nanoparticle/substrate interface. Our findings provide important insights towards the rational design of bimetallic plasmonic catalysts.

Reference: Sim, S.; Beierle, A.; Mantos, P.; McCrory, S.; Prasankumar, R. P.; Chowdhury, S., Ultrafast Relaxation Dynamics in Bimetallic Plasmonic Catalysts. *Nanoscale* **2020**.

597740 Combustion-Generated Carbonaceous Nanoparticles: The Impact of Oxidation on Their Morphology

Abstract:

Carbon black is the largest nanostructured material with annual production rate of just over 10 Mt/y worldwide. At the same time, soot emissions from fossil fuel and biomass combustion are comparable, just over 8 MT/y worldwide. As a result, there is renewed interest in understanding the detailed structure and optical properties of soot as it is the second largest contributor (after CO_2) to radiative forcing with the largest uncertainty of all such contributors. Furthermore, this understanding is still sought for reliable monitoring of carbon black manufacture and fire detection as well for mitigation of air quality impact on health and visibility reduction. Here, soot surface growth, agglomeration and oxidation are elucidated by discrete element modeling (DEM) and compared to experimental data at various combustion conditions from different laboratories. Emphasis now is placed on soot oxidation that takes place inside and on the surface of its constituent primary particles at a rate that depends on temperature, T, and O_2 concentration. Even though accurate oxidation kinetics are essential in both industrial uses and environmental impact of soot, they are often derived neglecting internal particle oxidation and the structure of such soot agglomerates. Here, the detailed evolution of the fractal-like agglomerate soot mass, m, and mobility diameter, d_m, during both internal and surface oxidation is determined by a moving sectional model. The model predictions are in excellent agreement with oxidation data of mature ethylene soot d_m for $T = 900 - 1200$ K. Then O_2 diffuses into the primary particles and reacts with bulk soot, hardly altering the d_m while at higher T, surface oxidation becomes dominant, decreasing d_m. The common assumption that soot agglomerates are spheres underestimates their d_m up to 50 % during oxidation. Coupling this detailed moving sectional model with soot mobility size distributions can yield realistic soot oxidation rates. Accounting for soot morphology and internal oxidation shows that the classic NSC rate increasingly underestimates (by 3 to 7 times) the oxidation rate of soot (from ethylene and toluene flames) with decreasing temperature (900-1800 K) and/or oxygen concentration (0.2 – 21 vol. %).

1. G.A. Kelesidis, S.E. Pratsinis, Internal and surface oxidation of soot agglomerates, *Combustion and Flame,* **209**, 493-499 (2019).
2. G.A. Kelesidis, M.R. Kholghy, J. Robertz, J. Zuercher, M. Allemann, A. Duric, S.E. Pratsinis, Light scattering from nanoparticle agglomerates, *Powder Technology*, **365,** 52–59 (2020).

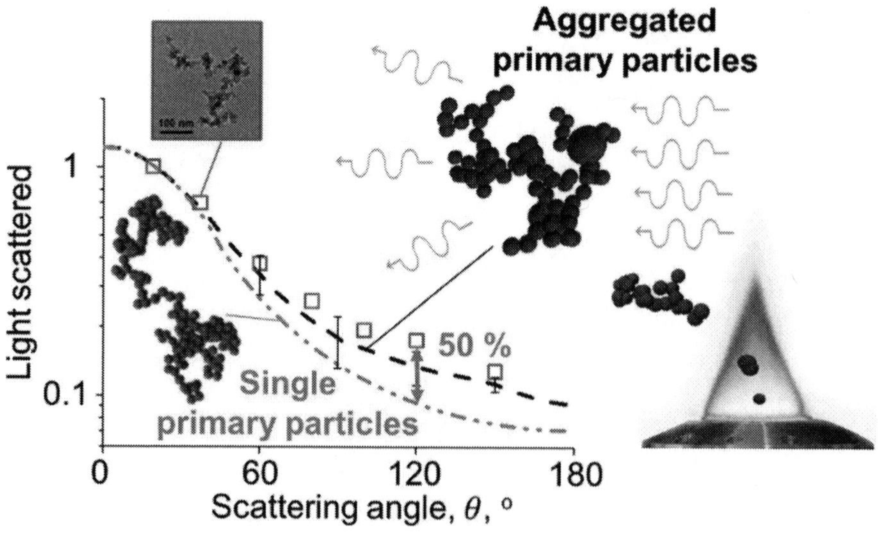

611332 Carbon Nanotube Photoluminescence in Chemical Biology

Abstract:

Area Plenary Invited Talks

Molecular probe development is a major pursuit of chemical biologists, to interrogate biological systems. Nanoscale tools, including probes and sensors, similarly enable new capabilities for biological investigation. The development of new fluorescent molecular sensors, by engineering triggered modulation of single-walled carbon nanotube photoluminescence, portends new capabilities outside of conventional chemical biology tools.

One form of carbon nanotube photoluminescence modulation, solvatochromism--the shifting of nanotube emission wavelength, can be induced via changes in the local dielectric/charge environment. We developed molecular triggers to modulate nanotube solvatochromic behavior to probe biological systems. We have developed solvatochromic sensors for nucleic acids, proteins, and metabolites, to enable the interrogation of living cells and animals.

Our recent investigations include temporal studies of signaling pathways, investigations of enzymatic inactivation, and the measurements of the accumulation of lipids specifically within the endolysosomal lumen to assess specific metabolic pathways in live cells and in vivo.

607187 Nanostructured Biomaterials for Immunomodulation and Cellular Therapy

Abstract:

A high level of spatial and temporal control over the presentation of biomolecules is essential for immune cell modulation. Nanostructured materials can be designed to provide efficient and tunable protein presentation. These structures can be tuned in terms of geometry allowing for greater residence time and modified to allow site-specific coupling to therapeutic proteins. In particular, nanowires and nanoparticles conjugated with antibodies can be designed to capture and potentiate endogenous cytokines demonstrating both tissue- and cell- specific immune activation. We show that the density and ratiometric control of immunomodulatory proteins such as costimulatory ligands and cytokines on can significantly impact human primary T cell activation phenotypes. This platform can be used to improve the safety of chimeric antigen receptor (CAR) T cell therapies by providing localized and specific cell activation for in vivo tumor clearance. Moveover, these tunable nanostructures can be used to modulate the microenvironment after injury, diminishing the fibrotic response. Our work highlights the role of tunable nanostructured biomaterials for immune modulation and cell based therapies.

596047 Plenary Talk: Nanomaterials Engineering to Probe and Control Living Systems

Abstract:

Unique physical, chemical, and optical phenomena arise when materials are confined to the nano-scale. We are accustomed to making observations and predictions for the behavior of living systems on a macroscopic scale that is intuitive for the time and size scales of our day-to-day lives. However, the building blocks of life: proteins, nucleic acids, and cells, occupy different spatiotemporal scales. Our lab focuses on understanding and exploiting tunable optical and mechanical properties of nanomaterials to access information about biological systems stored at the nano-scale. In the context of leveraging nanomaterial optical properties, we present recent work on developing and implementing neuromodulator nanosensors to image serotonin [1] or dopamine [2] volume transmission in the extracellular space of the brain, with implications for monitoring brain neurochemistry in neuropsychiatric and neurodegenerative disorders. We validate our dopamine nanosensor in acute striatal slices with electrical and optogenetic stimulation of dopamine release, and show disrupted dopamine kinetics when brain tissue is exposed to dopamine agonist or antagonist drugs [2,3]. In the context of leveraging nanomaterial chemical properties, we also discuss how high aspect ratio nanomaterials can be synthesized to carry biomolecular cargo to living systems. In particular, genetic engineering of plants is at the core of environmental sustainability efforts, but the physical barrier presented by the cell wall has limited the ease and throughput with which exogenous biomolecules can be delivered to plants. We will describe how nanomaterials engineering principles can be leveraged to genetically manipulate living plants [4,5], without transgene integration [6], in efforts to reconcile the benefits of crop genetic engineering with the demand for non-GMO foods [7]. Our work in the agricultural space provides a promising tool for species-independent, targeted, and passive delivery of genetic material, without transgene integration, into plant cells for rapid and parallelizable testing of plant genotype-phenotype relationships.

1. Jeong, S., Yang, D., Beyene, A.G., O'Donnell, J.T.D., Gest, A. M., Navarro, N., Sun, X., **Landry, M.P.**‡ High Throughput Evolution of Near Infrared Serotonin Nanosensors. *Science Advances* **5** (12), 1-12 (2019)
2. Beyene, A.B., McFarlane, I.R., Pinals, R.L, Landry, M.P.‡ Stochastic Simulation of Dopamine Neuromodulation for Implementation of Fluorescent Neurochemical Probes in the Striatal Extracellular Space. *ACS Chemical Neuroscience* **8** (10), 2275-2289 (2017).
3. Beyene, A. G., Delevich, K., O'Donnell, J.T.D., Piekarski, D.J., Lin, W.C., Thomas, A.W., Yang, S.J., Kosillo, P., Yang, D., Wilbrecht, L., **Landry, M.P.**‡ Imaging Striatal Dopamine Release Using a Non-Genetically Encoded Near-Infrared Fluorescent Catecholamine Nanosensor. *Science Advances* **5** (7), 1-11 (2019)
4. Demirer, G.S., Zhang, H., Matos, J., Goh, N., Cunningham, F.J., Sung, Y., Chang, R., Aditham, A.J., , Chio, L., Cho, M.J., Staskawicz, B., **Landry, M.P.**‡ High Aspect Ratio Nanomaterials Enable Delivery of Functional Genetic Material Without DNA Integration in Mature Plants. *Nature Nanotechnology* (2019). DOI: 10.1038/s41565-019-0382-5NNANO-18081684
5. Zhang, H.*, Demirer, G.S.*, Zhang, H., Ye, T., Goh, N.S., Aditham, A.J., Cunningham, F.J., Fan, C., **Landry, M.P.** Low-dimensional DNA Nanostructures Coordinate Gene Silencing in Mature Plants. *PNAS* (2019). DOI: 1073/pnas.1818290116
6. Demirer, G.S., Zhang, H., Goh, N.S., Grandio, E.G., **Landry, M.P.**‡ Carbon nanotube-mediated DNA delivery without transgene integration in intact plants. *Nature Protocols* (2019). DOI: 0.1038/s41596-019-0208-9
7. **Landry, M.P.**‡, Mitter, N.‡ How nanocarriers delivering cargoes in plants can change the GMO landscape. *Nature Nanotechnology* 2019, **14;** 512–514

606983 Multifunctional Nanoparticles for Imaging and Treating Traumatic Brain Injury

Abstract:

Traumatic brain injury (TBI) is the leading cause of morbidity and mortality in those under 44, yet there is no therapy that has improved outcomes in a large, multicenter Phase III trial. Adequate target engagement has been a major difficulty in novel treatment development, which reflects poor delivery into the brain. Multifunctional nanoparticles (NPs) offer the ability to simultaneously target affected tissue, track delivery in situ through imaging, and deliver treatments. With the goal of reducing the spread of secondary injury, we have developed oxygen reactive polymer, core crosslinked NPs that can promote accumulation and retention in a TBI, bind and inactivate reactive oxygen species in the thioether core, and be tracked through magnetic resonance imaging (MRI). In this work, we identify NP properties that promote accumulation and retention within and around injured brain and better understand a mechanism of therapeutic efficacy using a controlled cortical impact (CCI) mouse model of TBI. NP accumulation and retention kinetics were imaged using dynamic contrast enhanced MRI to determine the volume transfer coefficient, K^{trans}, through a variable flip angle method combined with diffusion MRI to assess tissue characteristics. Therapeutic efficacy of the NPs was determined though confocal imaging to observe neurodegeneration and neuroinflammation and using the Morris water maze to assess spatial learning and memory. NPs of various sizes and surface charges rapidly accumulated in the injury within 1 hr post-injury, negatively charged, 89 nm NPs were maximally retained when injected 3 hrs post-injury, and were retained for over 14 hrs in damaged brain as compared to the small molecule contrast agent Gd-DTPA, which was retained for only 4 hrs. NP accumulation correlated with regions of reduced diffusivity suggesting a mechanism of improved retention. During the chronic phase of the injury (1 month post-injury), CCI mice that received antioxidant NPs had better improved hippocampal function as compared to untreated CCI mice as determined by Morris water maze performance. This antioxidant NP-mediated preservation of hippocampal function correlated with reduced bilateral spread of neuroinflammation including neuron-astrocyte-microglia triads in the various subfields of the hippocampus. Our findings suggest NP-mediated delivery may provide a more effective approach for improving target engagement and reducing the spread of secondary injury following a TBI.

611233 Effects of Adsorbates on Thermal Transport in Metal-Organic Frameworks

Abstract:

Fast heat transfer during gas adsorption/separations in metal-organic frameworks (MOFs) plays a key role in improving their efficiency for gas storage/capture. Despite its importance, heat transfer in MOFs is not well understood, and whether it is helped or hindered by the presence of adsorbates is an open question. Our computational and experimental research indicate that depending on the nature of guest-framework interaction, adsorbates can increase or decrease thermal conductivity. The presence of physisorbed guests introduces new phonon scattering channels resulting from gas-framework collisions, which serve to reduce thermal conductivity. On the other hand, enhancement in thermal conductivity is observed for e-2–Zn_2(dobpdc) (e-2 = *N*-ethylethylenediamine; $dobpdc^{4-}$ = 4,4′-dioxidobiphenyl-3,3′-dicarboxylate) in the presence of chemisorbed CO_2.

599707 Flare Gas – Carbon Dioxide Assisted Synergistic Biomass Reforming for Hydrogen Rich Syngas Production on Self-Regenerable Carbon Nanofiber (CNF) Supported Fe – Mo_2c Catalyst

Abstract:

Biomass reforming for hydrogen rich syngas production was achieved through synergistic co-processing with flare gas and carbon dioxide over Fe-Mo_2C catalyst supported on carbon nanofiber (CNF). Methane – carbon dioxide activated biomass gasification can produce syngas with H_2:CO of 2 to 4 with H_2 product concentration ranging from 40 to 70 vol.%. Hydrogen rich syngas was generated over 15 cycles on Fe-Mo_2C catalyst at temperatures of 800°C, 850°C, and 900°C with varying Fe loading of 0.5 to 5 wt.% and Mo_2C concentration of 4 wt.%. CNF supported catalyst subjected to multiple cycles showed excellent resistance to deactivation and was discovered to be self-regenerable. The amount of iron loading was instrumental in the nature of support synthesized from pyrolysis at 700°C. Results showed that 0.5Fe catalyst led to synthesis of graphene supported nanoclusters of Fe and β-Mo_2C, whereas 2.5% Fe and 5% Fe loading materialized into CNF / CNT supported nanoparticles of Fe, β-Mo_2C. Conversion and gas yield nosedived to < 20% in 2 of 15 cycles on Fe-Mo_2C-CNF catalyst. These regeneration cycles possibly indicate nucleation or saturation of graphene layer synthesis or CNF growth. Nucleation of regeneration occurred mostly in cycle 2 and continued until saturated regeneration cycle varying in occurrence from subsequent cycles. Self-regeneration was the effect of free-C on the surface and complete conversion of Fe^0 active sites to Fe_3C / $Fe_3C_{1.5}$ active sites. Once elemental Fe was converted to Fe_3C_x, free-C led to 2D graphene layer formation on 2.5% Fe and 5% Fe catalysts. On graphene supported 0.5% Fe catalysts, CNF growth was observed on Fe_3C. Mo_2C active sites were much stable and did not participate in catalyst regeneration process. Self-regeneration mechanism for the CNF supported catalyst was proposed incorporating carbon from lignin oxygenates after being hydrodeoxygenated on β-Mo_2C in the framework of original mechanism from methane reforming perspective.

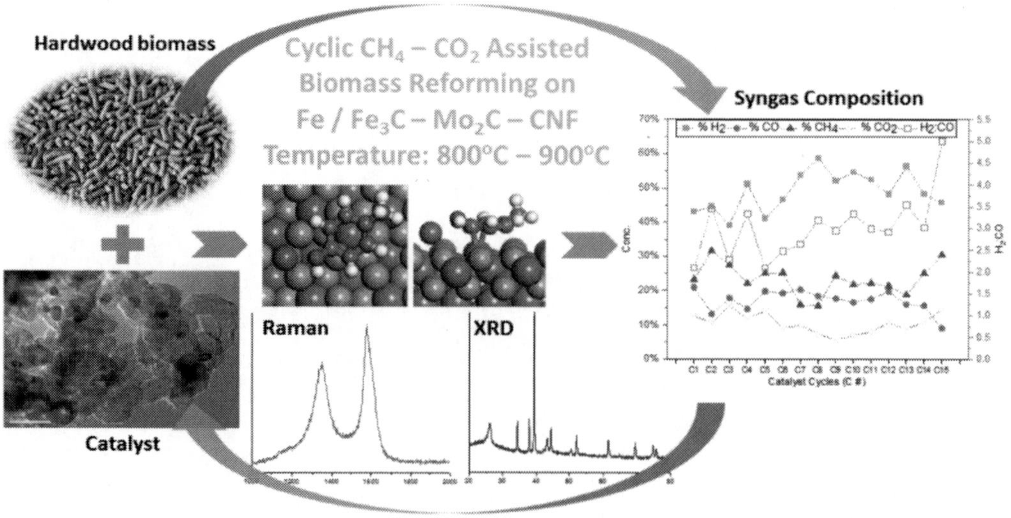

600613 A Sustainable and Scalable Technology to Produce High Performance Mesoporous Silicon for Lithium Ion Batteries

Abstract:

This presentation will provide an overview of our recent efforts on developing a new technology for the sustainable and scalable manufacturing of mesoporous silicon for the application of lithium ion batteries (LIBs). Silicon has the potential as an anode material to increase lithium-ion cell capacity but the associated volume change during lithiation leads to electrode degradation and loss of capacity during cycling. While mesoporous morphologies can potentially address these issues, magnesiothermic reduction (MgTR) of silica to silicon is currently the only method for potential bulk manufacturing. However, due to a distinct lack of mechanistic understanding of this process, predictive and sustainable design of mesoporous silicon is not possible [*J. Mater. Chem. A*, 2018, **6**, 18344].

To better understand the processes, we have systematically studied the process conditions using a model silica system [*J. Mater. Chem. A*, 2020, **8**, 4938]. Further, we performed a complete characterisation of both the reactant and product structures in order to map how reaction conditions affect the properties of the electrode materials and provide a processing-structure-property-performance relationship. Using this knowledge, we demonstrated the design of high performance silicon using any given silica feedstock.

Next, in order to make the process scalable and sustainable, we used three strategies.

(1) We used bioinspired silica (BIS) as a sustainable, economical and scalable feedstock [PCT/GB2016/052705] and the battery performance of porous silicon structures thus produced showed that they were superior compared to other silica sources.

(2) Decreasing the MgTR reaction temperature to make it more sustainable compromises the yield due to an insufficient drive to overcome the activation energy barrier. We addressed this challenge by reducing the particle size of the feedstock to the nanoscale such that their high surface energy could help overcome the energy barrier. We obtained high yields of porous silicon at temperatures previously shown to be too low for the MgTR to even proceed. This phenomenon can be exploited in order to trigger the conversion of larger, commercial sources of silica at low temperatures to porous silicon with high yields.

(3) Scaling up MgTR has proven difficult since, at large scales, the thermal effects of the MgTR are significant and lead to sintering and loss of porosity. By carefully investigating the role of NaCl as a thermal moderator, we identified the critical factors for controlling scale-up. Trials at different scales showed that applying this knowledge of the thermal moderators, we were able to maintain product properties at larger scales.

Overall, we report the first process for the predictive design of high-performance mesoporous silicon for LIBs such that the method is scalable and sustainable. Our results have established pathways for scaling-up MgTR method such that it can now be taken forward to target specific porous silicon properties.

607909 Modular Fluidic Microreactor for Fully Decoupled Precursor Mixing and Reaction Times in Mechanistic Studies of Metal Halide Perovskite Quantum Dot Synthesis

Abstract:

Metal halide perovskite quantum dots are a rapidly growing class of nanomaterials which have sparked significant interest for their broad reaching applications spanning from photocatalysis to optoelectronic devices.[1] Despite the high value of metal halide perovskite QDs, their synthesis optimization and control have been mostly limited to inefficient and inconsistent flask-based processes. Such batch synthesis reactors consume a large volume of often high-cost materials, require extensive manual labor, and possess inconsistent heat and mass transfer properties when moving between experiments and from flask to flask. These issues impair the accuracy and reduce the effectiveness of existing research efforts. In response, recently microscale flow synthesis strategies have been developed towards more efficient and systematic studies of perovskite quantum dots.[2–4] Among these flow synthesis platforms, fluoropolymer tubing-based microreactors are another quickly rising area of research, as they offer low-cost, fully modular, and commercially available microreactor components with quick assembly time as compared to conventional chip-based systems. These fluidic platforms offer micro-liter scale reactions volumes with automated, high-throughput, and non-invasive measurement strategies.[5]

However, as a property inherent to continuous flow processes, reaction time and mixing rates are strongly coupled to the total fluid velocity moving along the flow direction. In this work, we have developed a microscale fluidic microprocessor strategy which can independently control mixing times from 53 ms to 7.3 s with an arbitrarily long residence (*i.e.*, reaction) time. This microscale flow synthesis approach was accomplished by first developing and analyzing a fluoropolymer-based passive micromixer developed using commercially available tubing. We then integrate the passive micromixer module with an automated, material-efficient fluidic sampling module, which allows for two independent flow rates to run simultaneously within the platform. Utilizing the developed modular fluidic platform, "samples" with a specific chemical formulation are pulled from the mixing line and passed to the reaction/analysis line where they continue to react after forming homogenous reactive phase plugs. These plugs move through the flow reactor channel towards a spectral monitoring module for *in-situ* UV-Vis absorption and photoluminescence measurements.

In the next step, we applied the developed modular flow synthesis platform towards the room-temperature synthesis of cesium-lead-bromide perovskite quantum dots [6] and unveiled a mechanistic understanding of the quantum dot growth process through time-resolved spectral sampling. We then conducted a full compositional analysis coupled with variations in the precursor mixing time to attain highly tunable emission colors as well as quantum dot size distributions, quantum yields, reaction yields, and numbers of quantum dots formed. These experiments were coupled with a first principle mesoscopic kinetic Monte Carlo simulation to provide further insights into the mechanism behind the ligand-assisted reprecipitation process at early time scales. The strategies and models developed in this work offer a greater understanding of controlled synthesis of perovskite quantum dots with fast formation kinetics.

References

[1] Q. A. Akkerman, G. Rainò, M. V. Kovalenko, L. Manna, *Nat. Mater.* **2018**, *17*, 394.

[2] K. Abdel-Latif, R. W. Epps, C. B. Kerr, C. M. Papa, F. N. Castellano, M. Abolhasani, *Adv. Funct. Mater.* **2019**, *29*, 1900712.

[3] R. W. Epps, K. C. Felton, C. W. Coley, M. Abolhasani, *Lab Chip* **2017**, *17*, 4040.

[4] L. Bezinge, R. M. Maceiczyk, I. Lignos, M. V. Kovalenko, A. J. deMello, *ACS Appl. Mater. Interfaces* **2018**, *10*, 18869.

[5] S. Marre, K. F. Jensen, *Chem. Soc. Rev.* **2010**, *39*, 1183.

[6] S. Wei, Y. Yang, X. Kang, L. Wang, L. Huang, D. Pan, *Chem. Comm.* **2016**, *52*, 7265.

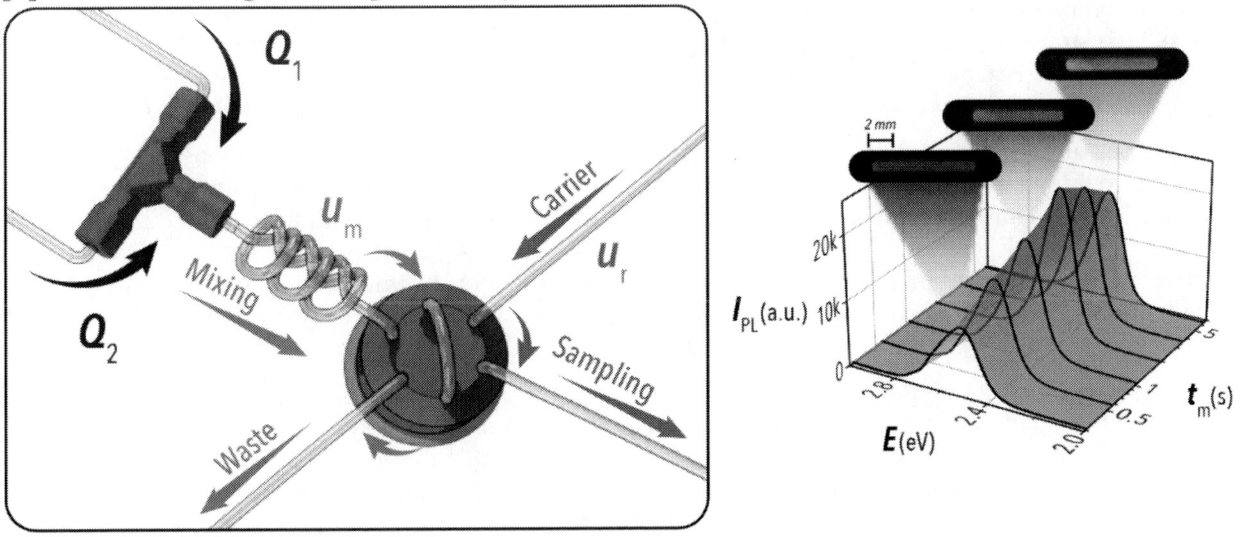

610022 Scalable Biomineralization of AgInZnS Quantum Dots for Photocatalytic Hydrogen Generation

Abstract:

Cadmium-based metal chalcogenide quantum dots are promising semiconductors capable of meeting the demands for photocatalytic H_2 generation, yet their intrinsic toxicity has motivated the search for high-performance, non-toxic alternatives.[1–3] The low toxicity, suitable band gaps, high extinction coefficients, and long photoluminescence lifetimes characteristic of ternary I-III-VI QDs makes them promising candidates especially when Zn is also incorporated to improve quantum yield (QY) and passivate the surface. One particularly promising composition is AgInZnS, owing to the tunability of its band gap (1.8-2.4 eV) to within the range suitable for photocatalytic H_2 generation.[4] However, the low compositional toxicity of AgInZnS is undercut by the high-temperatures and toxic solvents conventionally employed for its synthesis.[4,5]

Biomineralization-based routes to size-controlled QDs offer an alternative, 'green', and potentially scalable synthesis strategy, characterized by low temperatures and aqueous phase processing. Biomineralization of size-tunable CdS QDs, for example, is possible by the aqueous-phase, low-temperature enzymatic (*cystathionine gamma lyase*) turnover of the amino acid L-cysteine to H_2S in buffered solutions of cadmium acetate.[6] In this study, we expand the scalability and versatility of QD biomineralization by 1) establishing strategies for cyclic QD synthesis with facile enzyme recovery and reuse, and 2) elucidating mechanistics insight into single-enzyme synthesis of compositionally complex QDs. Specifically, we establish facile enzyme immobilization strategies on inexpensive, easily recoverable supports that offer sufficient enzyme activity to support continuous cyclic synthesis of QDs with narrow particle size distribution. This facile strategy for immobilized enzyme-mediated QD synthesis opens exciting possibilities for the scalable, 'green', biosynthesis of consistently sized QDs.

We then expand the compositional diversity of the single-enzyme biomineralization approach to AgInZnS QDs. In order to elucidate the AgInZnS formation mechanism, we carry out both one pot and sequential syntheses that identify critical In/Cys ratios affecting photoluminescence, offer insight into the role of Ag content in nucleation and growth, and begin to elucidate In-S complexation and the role of Ag ion exchange in tuning photoluminescence emission wavelength. We will demonstrate synthesis-structure-function relations in the context of photocatalytic hydrogen generation rates. Ultimately, this study advances the scalability and versatility of single-enzyme biomineralization as a means for 'green' synthesis of various non-toxic semiconductors.

References

1. Rao, V. N. *et al.* Sustainable hydrogen production for the greener environment by quantum dots-based efficient photocatalysts: A review. *Journal of Environmental Management* (2019) doi:10.1016/j.jenvman.2019.07.017.
2. Liu, Y. *et al.* One-step aqueous synthesis of highly luminescent hydrophilic AgInZnS quantum dots. *J. Lumin.*(2018) doi:10.1016/j.jlumin.2018.05.040.
3. Zang, Z. *et al.* Tunable photoluminescence of water-soluble AgInZnS–graphene oxide (GO) nanocomposites and their application in-vivo bioimaging. *Sensors Actuators, B Chem.* (2017) doi:10.1016/j.snb.2017.07.144.
4. Tang, X. *et al.* Nanocomposites of AgInZnS and graphene nanoshee photocatalysts for hydrogen evolution. *Nanoscale* (2015) doi:10.1039/c5nr05145b.
5. Lin, P. C., Wang, P. Y., Li, Y. Y., Hua, C. C. & Lee, T. C. Enhanced photocatalytic hydrogen production over In-rich (Ag-In-Zn)S particles. *Int. J. Hydrogen Energy* (2013) doi:10.1016/j.ijhydene.2013.04.125.
6. Spangler, L. C., Cline, J. P., Kiely, C. J. & McIntosh, S. Low temperature aqueous synthesis of size-controlled nanocrystals through size focusing: a quantum dot biomineralization case study. *Nanoscale* **10**, 20785–20795 (2018).

610583 Solution Behavior of Cellulose Nanocrystals in Polydisperse Polymer Solutions

Abstract:

The fabrication of cellulose nanocrystals (CNCs) into high-performance materials has gained significant attention as promising example of a renewable, sustainable, and inexpensive resource for advanced applications. Included in their innate "green" characteristics is a list of desirable chemical and mechanical properties: thermal stability, facile functionalization, and high theoretical tensile strength, that have been shown to translate well into macroscopic properties when used as a nanocomposite. All CNC composite materials require a "in-solution" processing step where solution interactions dictate the aggregation, interfacial, and liquid crystalline behavior. However, elucidating solution behavior is non-trivial and poorly reported for complex systems. We characterize the solution behavior of acetylated CNCs under varying solvent conditions with the addition of a polydisperse technical lignin. Lignin is a branching, heterogenous, biopolymer and provides useful insight into the robustness of CNC composite systems as we evaluate aggregation effects across the whole spectrum of loading conditions. Dilute solutions are characterized using Multi-Angle Light Scattering (MALS) and Dynamic Light Scattering (DLS) and it was observed that the addition of CNCs into lignin solutions improved dispersion and rheological properties for carbon fiber spinning. Included in the analysis are SANS experiments that show distinct conformational behavior for lignin as well as cross polarized light microscopy where we observe disperse solutions. Results show promise for using CNCs as a means for tuning complex phase behavior in multi-component polymeric solutions.

612871 Temporal Delivery of Polypeptide Nanoparticles for Accelerated Wound Healing and Tissue Repair (Award Session)

Abstract:

Complexity of repair process, low potency of monotherapies and delivery challenges limit the effectiveness of current wound healing treatment strategies, particularly in chronic and diabetic wounds. Growth factors, engineered cells and skin grafts have been investigated for promoting wound healing. However, little attention has been given to the critical role of augmenting early-stage processes in order to set the stage for effective execution of subsequent proliferation and remodeling phases of tissue repair. Here, we demonstrate that temporal delivery of a combination of bioactives that modulate and augment individual phases of healing leads to improved healing outcomes compared to sequential delivery. Histamine, a small-molecule chemoattractant of cells that mediate early stages of tissue repair, was delivered to full-thickness 5 mm excisional splinted wounds in the dorsal skin of immunocompetent BALB/c mice and genetically obese and diabetic db/db mice. Delivery of histamine along with silk fibroin dressing resulted in effective wound closure and improved healed skin strength in both Balb/c and db/db mice compared to Tegaderm dressing alone. Histological and immunohistochemistry analyses showed silk dressing-histamine treatment reduced dermal gap, promoted angiogenesis (CD31+), myofibroblast-mediated wound contraction (aSMA+), and higher TGF-β1 expression which are hallmarks of improved healing outcomes. Growth factors play key roles in different stages of tissue repair, especially in later stages i.e. proliferation, but their availability is limited due to dysfunctional healing response and proteolytic degradation in case of chronic wounds. We have used recombinant methods to generate growth factor nanoparticles (GFNPs) based on a growth factor-elastin-like polypeptide (GF-ELPs) fusion platform. Nanoparticles of basic fibroblast growth factor (bFGF-ELP) and stromal-derived growth factor 1 (SDF1-ELP), with hydrodynamic diameters from 300-400 nm, demonstrated higher stability and bioavailability in wound fluid compared to growth factors alone. Temporally sequential delivery of SDF1-ELP or bFGF-ELP nanoparticles immediately prior to the transition phase following histamine delivery resulted in improved biomechanical and biochemical outcomes of tissue repair compared to simultaneous delivery or individual treatments acting alone. These findings demonstrate that multifactor therapeutics (silk fibroin dressing, exogenous histamine, and GFNPs) are promising for improved outcomes in wound healing and tissue repair, including in diabetic and obese mice, and outperform clinically approved polyurethane wound dressing. A fundamental understanding of tissue repair kinetics and associated temporal delivery of therapeutics can lead to faster healing and improved tissue repair outcomes, thus reducing the propensity for re-opening, progression to intractable phenotypes and surgical site infections, all of which can have remarkable outcomes in the clinic.

612875 Microgliosis in a Dish (Award Session)

Abstract:

Microgliosis, due to brain injury or disease, can lead to varying molecular and morphological changes in microglia. Magnetic resonance elastography and ultrasound studies show that brain stiffness dynamically changes with age and disease state. Molecular mechanisms that account for the stiffness predilection to brain dysfunction, specifically on microglia, during brain disorders have been underexplored, and the impact of stiffness on microglia dysfunction and alterations during brain trauma is a gap in knowledge. The central goal of this study is to understand the role of mechanotransduction (stiffness) in microglia molecular and functional remodeling leading to increase brain dysfunction risk.

In this study, we have developed an innovative biomimetic platform named "BEASTS (Bio-Engineered Adhesive Siloxane substrate with Tunable Stiffness)" based on polydimethylsiloxane (PDMS) substrate in combination with our patented polyelectrolyte multilayer film (PEM)-coating technology to engineer mechanically tunable substrates to mimic physiologic (2 kPa),; early microgliosis (8 kPa); and diseased (25 kPa) stiffness to investigate the effect of stiffness on microglia activation and function. We observed that microglia grown on soft substrates (2 kPa) displayed a consistently more quiescent phenotype while those on stiff substrates (8 kPa and 25 kPa) displayed microgliosis-like morphology. In addition to morphological changes, microglia cultured on thick substrates demonstrated a significant increase in other microgliosis hallmarks- cellular proliferation and growth. Furthermore, culturing microglia on stiff surfaces resulted in increased reactive oxygen species (ROS) production. A similar effect has been observed in microglia in brain trauma animal models and patients. These data suggest a plausible mechanism that increased stiffness modulates microglia function, causing brain injury. Our platform lends itself to the study of potential therapeutic strategies for brain injury focusing on the intricate brain microenvironment-microglia signaling pathways.

612873 Near-Infrared Optical Detection of Dopamine Based on Xeno Nucleic Acid (XNA) Sensors (Award Session)

Abstract:

Single-walled carbon nanotubes (SWCNTs) emit near-infrared fluorescence that is beneficial for in vitro and in vivo optical sensing. State-of-the-art SWCNT sensors are functionalised with single-stranded DNA (ssDNA) sequences that control the dispersivity, brightness, biocompatibility, and selectivity of the sensor. However, the fluorescence of the ssDNA-functionalized nanotube sensors is sensitive to changes in the local ion concentration. This sensitivity is problematic not only for measurements in complex biofluids, but also for sensing neurotransmitters, whose concentrations are strongly coupled to spatial and temporal ion fluctuations. The fluctuating ion concentrations can therefore interfere with the accurate optical detection of neurotransmitters, thereby limiting the use of ssDNA-SWCNT sensors for detecting and treating neurological diseases.

We present an optical SWCNT dopamine sensor with improved stability towards cation-induced fluorescence changes. The synthetic biology approach presented in this work serves as a complementary means for enhancing nanotube optoelectronic behaviour, unlocking previously unexplored possibilities for developing nano-bioengineered sensors with augmented capabilities.

612865 Inert Nanoparticles for Enhancing the Survival of Primary Macrophages (Award Session)

Abstract:

Cell-based therapies have recently emerged as a new platform for treating a spectrum of chronic degenerative diseases, with serious advances made especially in hematopoietic stem cell and reactive lymphocyte therapies. Macrophages are leukocytes responsible for phagocytosing foreign objects, bacteria, and apoptotic cells, maintaining homeostasis, and bridging the innate and adaptive immunity. Macrophage-based cell therapy has been identified as a promising therapeutic approach to treat chronic immune dysfunctions, owing to the functionality and phenotype plasticity of macrophages. However, a major challenge in the advancement of macrophage-based cell therapies is the difficulty to maintain adequate cell viability *ex vivo* following macrophage isolation from tissue. Current attempts of macrophage *in situ* activation or autologous cell therapy following *ex vivo* stimulation fail to elicit effective therapeutic responses, due to low cell survival rates upon reintroduction. Therefore, there is a need to develop new approaches to regulate primary macrophage survival in both *in vivo* and *ex vivo* environments.

We attempted to extend the viability of *ex vivo* macrophage cultures through treatment of inert poly(ethylene glycol) (PEG) diacrylate (PEGDA)-based nanoparticles. We observe that *ex vivo* bone marrow-derived macrophages (BMMs), which typically undergo apoptosis within 2 weeks under appropriate culture conditions, can persist and maintain functionality for up to 6 months after being treated with a *single* dose of PEGDA nanoparticles. We find that nanoparticle uptake significantly delays primary macrophage cell death through the downregulation of caspase-dependent apoptosis pathways. Furthermore, we observe no major changes to macrophage phenotype following nanoparticle uptake. However, phagocytosis of PEGDA nanoparticles leads to increased expression of major histocompatibility complex (MHC) class II uniquely in the absence of other co-stimulatory molecules. The enhanced survival phenomenon is also applicable to *ex vivo* cultures of terminally differentiated macrophages isolated from tissue, including peritoneal and alveolar primary cells. Overall, this work demonstrates for the first time the ability of inert nanoparticles to suppress apoptotic signaling and prolong viability in primary macrophages extracted from different tissues. This work could eliminate a major obstacle standing in the way of developing macrophage-based cell therapies, as well as provide mechanistic insight into the implications of nanoparticle phagocytosis on cell longevity and function.

612872 Fusogenic Liposomes Loaded with Antibiotics As a Combined Therapy for Fighting Brain Infections (Award Session)

Abstract:

Meningitis is caused by the inflammation of meninges and the in-between fluid filled spaces. Bacterial meningitis, the most common and the most aggressive form of bacterial infection, is mainly caused by Gram-positive bacterium *Streptococcus pneumoniae* and Gram-negative bacterium *Neisseria meningitidis*. Its treatment consists of a combination of antibiotics, that sometimes fail due to their poor permeability through the blood-brain barrier (BBB). Moreover, the emergence of drug-resistant bacteria and their ability to create biofilms have become a global concern, making it increasingly difficult to treat these infections. When bacterial biofilms are formed, the ability of antibiotics to treat these diseases is diminished, leading to ineffective treatment. In severe cases, meningitis can lead to death. Therefore, there is a need to create alternative treatments for bacterial infections of brain, such as bacterial meningitis.

The use of nanotechnology in medicine, especially in drug delivery, is one of the most promising strategies to solve the problem of antibiotic resistance. Here, we report the use of fusogenic (pH-sensitive) liposomes, as drug delivery systems to target the bacteria cell membrane. These liposomes release the antibiotic directly at the site of infection, notably inside the bacterial cell. pH-sensitive liposomes have been previously used to target bacteria due to their ability to interact with the phospholipids present in the bacteria outer membrane, destabilizing it. For the purpose of this study, liposomes were functionalized with a cell-penetrating peptide, TAT (47-57) to increase their permeability through the blood-brain barrier, which is the primary challenge in treating meningitis.

Lipid film rehydration method was used to synthesize the liposomes, followed by their functionalization with TAT (47-57) peptide using wet chemistry. Liposomes were characterized by transmission electron microscopy (TEM) to verify their formation, and size of about 100 nm. These liposomes were loaded with three different antibiotics, methicillin, ampicillin and methicillin, with a loading efficiency of around 68%, 72% and 83% respectively.

Bacterial growth curve studies, and colony forming studies results showed that TAT functionalized liposomes loaded with vancomycin and methicillin were successfully able to kill *Streptococcus pneumoniae*, and methicillin-resistant *Staphylococcus aureus* respectively. These results not only demonstrate the ability of these particles to decrease the minimum inhibitory concentration (MIC) of antibiotic necessary to kill bacteria, but also their ability to combat drug-resistant bacteria.

In vitro cytotoxicity test using fusogenic liposomes against astrocytes and endothelial cells, primary cells that form the BBB, showed no interaction between our nanoparticles and both cell lines, as well as a significant reduction of antibiotic cytotoxicity through encapsulation.

A novel static *in vitro* BBB model was used to determine the ability of these nanoparticles to cross to blood brain barrier. This model consisted of astrocytes seeded at the bottom of a 24-well plate, with inserts containing endothelial cells added to each well. After a set incubation time, endothelial cells were placed inside a separate 24-well plate containing bacteria. Liposomes were added to study their permeability through the endothelial cell barrier to reach the bacteria on the other side of the membrane. Initial studies with these static models showed the ability of our functionalized nanoparticles to not just cross the blood brain barrier, but also significantly decrease the bacteria population on the other site.

Future studies will be focused on a better understanding of the interaction mechanism between our TAT functionalized liposomes and bacterial cells, and on the use of dynamic and more sophisticated BBB models to better mimic the liposome permeability through the BBB.

This work was supported by Department of Chemical Engineering, Northeastern University.

612870 Photo-Crosslinked Nanogel *Via* Nanoreactor for Therapeutic Proteins Delivery (Award Session)

Abstract:

Nanotechnology-based drug delivery systems use a variety of nanoparticle fabrication methods to formulate the different types of delivery formula. The nano-delivery systems provide the enhanced membrane permeability of the drugs because the nanoscale delivery formula allows the maximum surface area of a particle per unit of mass.

In this study, the polymeric nano-hydrogel (Nanogel) was formulated to deliver therapeutic proteins. The material biocompatibility and delivery efficiency were first taken into consideration to make an effective delivery carrier. Nanogel was synthesized by a process of confining hydrophilic biopolymer in a water-in-oil nanoreactor followed by photopolymerization and purification. This system enhanced the penetration of fluorescent-conjugated bovine serum albumin into a biological membrane indicating its suitability for transdermal drug delivery [1]. The modified nanogel was formulated by using fish gelatin methacrylate to enhance the drug diffusion at the target location without significant cytotoxicity [2]. For testing delivery efficiency and therapeutic effect, a model therapeutic protein was synthesized using a cell-free system described in the previous work [3]. The human skin protein was then encapsulated in the drug carrier and tested on the skin membrane model. Nanogel system showed great potential in transdermal drug delivery of protein therapeutics.

[1] *Macromol Res* 24 (12), 1115

[2] *Mar Drugs* 17 (4), 246

[3] *Method Protoc* 2 (3), 68

612869 A Wearable Microfiber Biomaterial Incorporates Optical Nanosensors and Enables Wireless Monitoring of Oxidative Stress (Award Session)

Abstract:

Reactive oxygen species (ROS) are a family of signaling molecules continuously generated, transformed, and consumed by all aerobic organisms. Wound healing is one of the most dynamic biological process featuring ROS-linked cellular signaling throughout the entire process. Additionally, basal concentrations of ROS aid in the fight against invading microorganisms into open wounds. The excessive and uncontrolled production of ROS contributes to the sustaining and deregulation of inflammation processes, which play a central role in the pathogenesis of chronic non-healing wounds. Physiologically, hydrogen peroxide (referred herein as peroxide) and superoxide serve as intracellular ROS messengers stimulating key phases of wound healing including cell recruitment, production of cytokines, and angiogenesis. Of note, peroxide acts as the principal secondary messenger in wound healing and is present at low concentrations (100–250 μM) in normal wounds. Increased peroxide concentration is a biomarker for chronic wounds in which biofilm-making pathogens can grow significantly faster than normal acute wounds. Due to the short half-life of peroxide, detection and quantification of their concentrations are difficult in vivo and in patients. Thus, peroxide levels in wounds have generally been determined indirectly through analysis of downstream oxidation products of lipids, proteins, or DNA.

Single-Walled Carbon Nanotubes (SWCNTs) with engineered wrappings have recently been developed and utilized in various disparate fields ranging from additives that strengthen material composites to biomedical applications including targeted drug delivery, optical biosensing, and biological imaging. Of significant interest, SWCNTs exhibit intrinsic photoluminescence (fluorescence) that is photostable, tunable, and sensitive to its local environment. The electronic band gap energies of SWCNTs are dependent on their chiral identity and are differentially and reversibly oxidized, enabling the ratiometric detection of local redox environments.

With revolutionary advances in nanotechnology and biomaterials in the recent years, an extensive range of smart wound care biomaterials have been developed which incorporate components such as growth factors and/or antimicrobial agents such as iodine or silver. Electrospun micro/nanofibers are one of the novel classes of wound dressings as they mimic the chemical and mechanical environment of the 3D extracellular matrix. Micro/nanofiber-based wound dressings have been designed to enhance cell migration, prevent inflammation and infection and inhibit scar formation on wounds. Herein, we developed a process to construct wearable textiles incorporating the DNA-wrapped SWCNT nanosensors through a one-step co-axial electrospinning method. We then developed an optical microfibrous biomaterial with a "core-shell" morphology that encapsulates fluorescent SWCNT nanosensors. We chose Polycaprolactone (PCL) as the shell material as it is an FDA-approved polymer which has been extensively studied for tissue engineering and wound healing applications. Our optical biomaterial was able to ratiometrically detect peroxide in the wound biological range (20-500 uM). The ratiometric characteristic of the sensor will significantly facilitate the future in vivo and clinical applications as it transduces an absolute signal which is not dependent on excitation source distance and exposure time. Moreover, the optical nature of the biomaterial will enable the wireless detection without a need to disturb the wound dressing. We revealed that the signal from the biomaterial does not significantly decay over time as the fluorescence of SWCNTs is non-photobleachable. We also quantified the absolute weight of SWCNT nanosenors in the biomaterial over time using confocal -Raman microscopy and indicated that the nanosensors do not leak out of the fibers over time, mitigating any toxicological concerns related to SWCNTs. Our optical biomaterial was integrated into other common wound dressing materials such as hydrogels and commercialized bandages, showing the compatibility of this platform with other existing wound dressings. Thus, our wearable microfiber biomaterial enables wireless monitoring of oxidative stress directly from the wound site.

612874 Understanding Corona Exchange Dynamics on Carbon Nanotubes with Multiplexed Fluorescence Monitoring (Award Session)

Abstract:

Adsorption of polymers on single-walled carbon nanotubes (SWCNTs) has enabled developments in molecular sensing, *in vivo* imaging, and gene delivery applications.[1,2] Noncovalent functionalization of SWCNTs with polymers preserves the pristine SWCNT atomic structure, thus retaining the intrinsic near-infrared fluorescence for sensing or imaging functions, and offers a reversible binding mode for cargo delivery. However, noncovalent biomolecule adsorption is an inherently dynamic process, where exchange occurs between molecules in the bulk solution and molecules on the nanoparticle surface, into what is known as the corona phase. The nature, strength, and kinetics of both the polymer binding and unbinding processes on SWCNTs are important contributors to the success of polymer-SWCNT based technologies. Understanding this binding process is especially important for intended uses of functionalized SWCNTs in biological environments, where native biomolecules compete with the original polymer to occupy the SWCNT surface. Binding of proteins and other biomolecules to the SWCNT disrupts the intended functionality of the nanoparticle and can further lead to adverse biocompatibility outcomes.

We present an assay to study the corona exchange dynamics between solution-phase and corona-phase polymers on SWCNTs.[3] This assay exploits the quenching property of fluorophores proximal to the SWCNT surface to monitor ligand binding and unbinding events. Real-time tracking of polymer adsorption and desorption events on the SWCNT surface are conducted with systematic variation of the molecular entities and solution conditions, with a specific focus on DNA (as the sensing moiety) and protein (as the biofouling agent). Binding profiles are extracted from the experimental assay and used to inform a kinetic model of the system. The work presented herein develops an understanding of the fundamental corona exchange mechanism and provides insight into performance of these designed SWCNT-based systems in biologically relevant, protein-rich conditions.

References

1. Demirer, G. S. *et al.* High aspect ratio nanomaterials enable delivery of functional genetic material without DNA integration in mature plants. *Nature Nanotechnology* 1 (2019) doi:10.1038/s41565-019-0382-5.
2. Beyene, A. G. *et al.* Imaging striatal dopamine release using a nongenetically encoded near infrared fluorescent catecholamine nanosensor. *Science Advances* **5**, eaaw3108 (2019).
3. Pinals, R. L., Yang, D., Lui, A., Cao, W. & Landry, M. P. Corona Exchange Dynamics on Carbon Nanotubes by Multiplexed Fluorescence Monitoring. *J. Am. Chem. Soc.* (2019) doi:10.1021/jacs.9b09617.

605170 Interactions of Graphene Oxide Nanosheets with Blood-Related Entities and Their Implications for Hematological Disorders

Abstract:

Blood-related disorders are some of the most fundamental and complex abnormalities as they can cause critical health conditions, including heart attack, stroke, and cancer [1, 2]. Although nanomaterials have been actively explored for various biomedical applications [3-5], they have not been really investigated for hematological applications. Among all classes of nanomaterials explored for biomedicine, the two-dimensional (2D) graphene oxide (GO) nanosheets stand out because of their unique and versatile physicochemical properties [6], which may be beneficial for hematological applications [7-10]. Herein, we examined the interactions of GO nanosheets with different blood-related entities, particularly plasma proteins, blood cells, and blood-borne parasites, to unravel the biophysicochemical activities of GO nanosheets for hematological applications. First, we evaluated the GO-protein associations, specifically the adsorption, equilibrium binding, and conformational stability of plasma proteins upon interacting with GO nanosheets [7, 8]. The interactions of GO nanosheets with whole blood under flow condition, and with *Plasmodium falciparum* malaria parasites, were then probed to assess the antithrombotic and antimalarial properties of GO nanosheets [9, 10]. GO nanosheets were noted to possess high loading capacity for major blood plasma proteins and the nanosheets did not induce protein denaturation. These interactions were protein-specific and might be considerably influenced by the lateral size distribution and concentration of GO nanosheets. We also observed that functionalized GO nanosheets displayed antithrombotic characteristic under flow condition. Moreover, pristine GO nanosheets, irrespective of their size distribution, exhibited antimalarial activity by mitigating the malaria parasite invasion on red blood cells and delaying the parasite maturation. Overall, with their high biomolecule loading capacity and unique antithrombotic and antimalarial characteristics, GO nanosheets are highly promising for hematological applications. This study will provide a deeper insight into nanomaterial-blood interactions and potentially facilitate further explorations into the formulation of nanotechnology-based strategies for biomedical applications.

References

[1] Furie et al., *The New England Journal of Medicine* 359 (2008), 938-949.
[2] Burns et al., *The Lancet Haematology* 3 (2016), E362-E370.
[3] Doane et al., *Chemical Society Reviews* 41 (2012), 2885-2911.
[4] Chen et al., *Chemical Reviews* 116 (2016), 2826-2885.
[5] Shi et al., *Nature Reviews Cancer* 17 (2017), 20-37.
[6] Kenry et al., *Biomaterials* 155 (2018), 236-250.
[7] Kenry et al., *NPG Asia Materials* 9 (2017), e422.
[8] Kenry et al., *Nanoscale* 8 (2016), 9425-9441.
[9] Kenry et al., *Small* 11 (2015), 5105-5117.
[10] Kenry et al., *Nanoscale* 9 (2017), 14065-14073.

605188 A Wavelength-Induced Frequency Filtering Method for Extending Fluorescent Biological Assays *In Vivo*

Abstract:

Fluorescence-based assays are employed ubiquitously throughout the life sciences and medicine, however their adaptation and translation to the *in vivo* environment is fundamentally hampered by unfavorable tissue scattering and intrinsic autofluorescence. Herein, we develop a wavelength-induced frequency filtering (WIFF) and show that it significantly improves the *in vivo* signal to noise ratio (SNR) of fluorescent probe and sensor components, extending measurements to extremely deep implants up to 5.5±0.1 cm deep in highly scattering tissue phantoms, complex tissues, and SKH1-E mouse models. By modulating the excitation wavelength cyclically across the absorption cross section of any sensing chromophore, the emission waveform can be deconvoluted from the scattering background, autofluorescence and other noise sources. The technique is shown to boost SNR up to 52-fold for ten common fluorescent probes used in biomedical assays and emit across the entire visible range. Near-infrared fluorescent carbon nanotube sensors are also probed with this technique to detect H_2O_2, riboflavin, and ascorbic acid at high fidelity *in vivo* with 2 Hz temporal resolution in real time. These results based on WIFF open up new avenues of biomedical research by extending a large number of fluorescent sensing assays to previously inaccessible *in vivo* environments.

611366 Self-Assembled Rosettes As Chemically Defined Corona Phases for One Dimensional Nanomaterials

Abstract:

Liquid phase dispersion and exfoliation is one of the primary methods of large batch production of singly separated nanomaterials, which is often a step toward harnessing the unique properties attributed to materials at the nanoscale. In the case of aqueous solutions, hydrophobic materials such as graphene or single-walled carbon nanotubes (SWCNTs) necessitate either chemical functionalization or non-covalent association with amphiphilic components to remain stable during experimental timescales. In the case of non-covalent conjugation, the surface arrangements of the dispersants often create an amorphous corona phase around the nanoparticle, complicating the material control and design. In this work, we propose the use of self-assembled nanostructures consisting of heteroaromatic bicyclic bases, rosette nanotubes (RNTs), to encapsulate SWCNTs in the dispersion process. We demonstrate and characterize singly-encapsulated RNT-SWCNT complexes that has a much higher surface coverage and water exclusion than most other aqueous SWCNT systems. In addition to a SWCNT diameter preference in the dispersion process, the encapsulated SWCNT shows both diameter and chirality dependent photophysical phenomena that is linked to the RNT structure. Thus, the RNT-SWCNT system acts as a regularly defined corona phase affords unprecedented control of SWCNT properties and subsequently function.

609319 Synergistic Effects of Carbohydrate Identity and Polymer Chain Length on Creating Glycopolymer-Wrapped Carbon Nanotubes

Abstract:

Dispersion of single-wall carbon nanotubes (SWCNTs) with biopolymers in aqueous environments facilitates their potential biological and biomedical applications. Glycopolymers, biomimetic polymers with carbohydrate pendant groups, have great potential for dispersion of SWCNTs since they have greater structural tunability and targeted biological functions compared to natural polysaccharides. In this work, we investigated a small library of synthetic glycopolymers with monosaccharide and disaccharide groups for dispersion of SWCNTs *via* non-covalent complexation in aqueous condition. We found that glycopolymers can effectively stabilize SWCNTs to form water-soluble glycopolymer-wrapped SWCNT complexes, depending on the carbohydrate density and polymer chain length. This work demonstrates for the first time the synergistic effects of the identity of carbohydrate pendant groups and polymer chain length of glycopolymers on stabilizing SWCNTs in water.

608453 An Antibiotic-Free Nano-Enabled Approach for the Topical Eradication of Dental Biofilm without Disturbing Microbiota Balance *In Vivo* Using Carbon Dots

Abstract:

Dental plaques are one of the most prevalent types of biofilm which can lead to dental caries due to the demineralization processes caused by acidogenic bacteria. These bacteria reside inside a protective sheath *i.e.* extracellular polymeric substances which makes any curative biofilm treatment challenging. Herein, we propose an antibiotic-free strategy to disrupt the biofilm by utilizing nanoparticles designed to function in the acidic environment of these biofilms. Carbon dot nanoparticles were decorated with phosphonium ions to endow them with antibiofilm properties and they were made stimuli-responsive by their entrapment in a pH-sensitive polymer shell. Mechanistic studies on *Streptococcus mutans* (*S. mutans*) revealed that the toxicity of these nanoparticles was mediated by their electrostatic interactions with the bacterial membrane, the generation of excess reactive oxygen species (ROS), and subsequent DNA fragmentation. *In vitro* and *Ex vivo* studies on mature biofilms revealed more than 90% biofilm inhibition. An *in vivo* examination in a rat model of dental biofilm further showed that the nanoparticles could effectively suppress the growth of *S. mutans*. The nanoparticles were biodegradable in artificial saliva after ten days and did not show any adverse effects on the surrounding oral tissues or on internal organs. Importantly, 16S rRNA analysis of the dental microbiota showed that the diversity and richness of bacterial species did not significantly change with nanoparticle treatment. Overall, this study presents a safe and effective, antibiotic-free approach to decrease the *S. mutans* population and dental biofilm formation without disrupting the ecological balance of the oral cavity.

604293 Observation of Nonlinear Optical Effects from Carbon Dots and Applications for Fluorescence Bioimaging from Cells through Tissues to Animals

Abstract:

Carbon dots (CDs), also known as carbon quantum dots or carbon nanodots, generally refer to small carbon nanoparticles with a typical size under 10 nm in all three dimensions, which have found great potential for a wide range of applications, from chemical sensing through bioimaging to optoelectronics. As for fluorescence bioimaging, the penetration of visible light into biological tissues is quite limited because of the large photon absorption and scattering, which results in a distinct photon attenuation and is unfavourable for deep-tissue fluorescence imaging. In this presentation, I will introduce our work on rapid and large-scale production of CDs, followed by the observation of their nonlinear fluorescence behaviors under excitation of femtosecond (fs) laser in both traditional near-infrared spectral region (NIR-I, 700–900 nm) and second near-infrared region (NIR-II, 1000–1700 nm). The two-photon excited fluorescence spectrum of the CDs by a 750 nm fs laser was observed in the wavelength range of 400–600 nm, with peak maximum located at ~460 nm and long fluorescence lifetime of ~7.2 ns, which were similar to the counterpart under one-photon excitation. However, the wavelength-tunable two-photon fluorescence from the CDs could be obtained with spectral peaks from 580 nm to 740 nm, under 1000–1560 nm fs laser excitation. Interestingly, when an fs excitation wavelength as long as 1560 nm was adopted, two typical nonlinear optical signals, namely two-photon fluorescence and third harmonic generation (THG), could be observed. Based on the nonlinear optical effects of CDs under NIR fs laser excitation, CDs-assisted two-photon fluorescence microscopy realized high-contrast fluorescence lifetime imaging of cancer cells, deep-tissue imaging of zebrafish embryos and the brain neuron networks of mice. These studies show the potential of CDs as fluorescent probes for deep-tissue and high-contrast functional bioimaging and related applications in the future.

604900 Templating Colloidal Sieves for Tuning Nanotube Surface Interactions and Optical Sensor Responses

Abstract:

Surfactants offer a tunable approach for modulating the exposed surface area of a nanoparticle. They further present a scalable and cost-effective means for suspending single-walled carbon nanotubes (SWCNTs), which have demonstrated practical use as fluorescence sensors. Though surfactant suspensions show record quantum yields for SWCNTs in aqueous solutions, they lack the selectivity that is vital for optical sensing. We present a new method for controlling the selectivity of optical SWCNT sensors through colloidal templating of the exposed surface area. Colloidal nanotube sensors were obtained using various concentrations of sodium cholate, and their performances were compared to DNA-SWCNT optical sensors. Sensor responses were measured against a library of bioanalytes, including neurotransmitters, amino acids, and sugars. We report an intensity response towards dopamine and serotonin for all sodium cholate-suspended SWCNT concentrations. We further identify a selective, 14.1 nm and 10.3 nm wavelength red-shifting response to serotonin for SWCNTs suspended in 1.5 and 0.5 mM sodium cholate, respectively. Through controlled, adsorption-based tuning of the nanotube surface, this study demonstrates the applicability of sub-critical colloidal suspensions to achieve selectivities exceeding those previously reported for DNA-SWCNT sensors.

602591 Synthesis, Characterization, and Functionalization of Graphene Oxide-Based Nanoplatforms for Gene Delivery

Abstract:

Gene therapy has been proposed as a promising strategy for the treatment of various genetic disorders and acute acquired diseases such as autosomal or X-linked recessive single-gene disorders including Cystic Fibrosis (C.F.), Adenosine Deaminase deficiency (ADA), emphysema, retinitis pigmentosa, sickle cell anemia, phenylketonuria, hemophilia, and Duchenne Muscular Dystrophy (DMD) [1]. Additionally, recent reports suggest potential treatments for some autosomal dominant disorders, polygenic disorders, different forms of cancers, vascular disease, neurodegenerative disorders, and inflammatory conditions, among many others [2]. Despite the success both *in vivo* and *in vitro*, one of the major challenges that still require much effort is the efficient delivery of the therapies, as only a fraction of the delivered genetic material effectively reaches the target site [3]. To address this issue some of the current approaches include the use of viral and non-viral vectors as delivery carriers [2, 3]. Viral vectors have been reported to deliver the gene cargoes with superior efficiencies but thus far have failed to comply with all regulatory and biosafety requirements [4]. These limitations have spurred the development of non-viral safer alternatives that induce a lower immune response and can be prepared inexpensively [2]. As a result, over the past few years, non-viral carriers have gained considerable attention, and particularly those based on nanostructured materials [2, 5]. In this regard, there is a large number of available nanomaterials with unique properties that could potentially be utilized as carriers of biological material [5]. Thus far, only a relatively small fraction of them has been tested for delivery purposes due to issues regarding the required complex synthesis and purification protocols, expensive reagents and most importantly, their limited ability to escape endosomes. These are compartments produced by the invagination of the cell membrane, which eventually matures into lysosomes where enzymes break down their contents [6]. This cell protection mechanism has become a significant obstacle for the full implementation of nanovehicles for drug delivery. This is mainly because the costly transported therapeutic molecules remain trapped and eventually are degraded and expelled from the cells [7]. To address this issue, the nanomaterials have been functionalized with cell-penetrating peptides and proteins, which not only help to bypass the cell membrane but in some cases allow endosome escape [6, 8]. In recent contributions, we have interfaced magnetite nanoparticles with the antimicrobial peptide Buforin II and discovered that the obtained nanobioconjugates exhibited potent translocating and endosomal escape abilities in several cell lines [9]. Despite the success, we wanted to extend these capabilities to other families of nanomaterials such that a more ample arsenal of platforms is available for the rational design of cell-penetrating vehicles to address the needs of particular tissues and diseases. A family of nanomaterials with unique physicochemical properties are the ones based on Carbon [10, 11]. In particular, graphene oxide (GO) has recently emerged as an alternative of great interest for gene and drug delivery mainly because of its unique 2D topology, high surface area, and electrical and optical properties [12]. Additionally, GO is relatively easy to synthesize with controlled particle size and new prospects for large-scale production are becoming available almost on a daily basis [13, 14].

Accordingly, in this work we propose to put forward a new family of cell-penetrating based on GO-based nanoplatforms by interfacing GO with potent translocating peptides. This approach will take advantage of already tested and new peptides as well as the unique interactions of GO with the phospholipids of membranes and endosomes. Ideally, the platform should be able to carry a plethora of pharmacological agents but we are particularly interested in delivering gene therapies. GO was synthesized by the modified Hummers' method through the oxidation of graphite sheets [6]. Briefly, 0.75 g of graphite sheets were mixed with 4.5 g of $KMnO4$ in a solution of $H2SO4$ (90 mL) and $H3PO4$ (10 mL). The mixture was left at 50°C and under continuous stirring at 350 RPM for 12 hours. Next, the mixture was cooled down at 0°C and $H2O2$ was subsequently added to quench the oxidation reaction. The resulting mixture was then exfoliated to obtain a solid that was centrifuged followed by resuspension in an aqueous solution of HCl and ethanol. Finally, the as-synthesized GO was filtered, lyophilized, and stored at 4°C until further use. 100 mg of an aqueous suspension of GO were then silanized by adding 2 mL of tetramethylammonium hydroxide (TMAH) 25% (v/v), 50 µL of pure acetic acid and 1 mL of (3-Aminopropyl) triethoxysilane (APTES) 10% (v/v). A scheme of the reaction is shown in Fig 1A. The silanization was to render free amine groups on the surface of GO to further conjugate the peptide molecules. Thermogravimetric analysis (TGA) and Fourier-Transform Infrared spectroscopy were used to characterize the nanoplatform. FTIR analysis showed the distinctive peaks related to the characteristic carboxyl groups of GO (Fig 1B) as well as the Si-O bonds after silanization (Fig 1B). TGA allowed us to estimate a silanization efficiency of about 35%.

Future work will be focused on conjugating Buforin II and evaluating translocation efficiency by conducting uptake experiments in liposomes and various cell lines. Additionally, we will determine endosomal escape via confocal microscopy by labeling the peptide with fluorescent molecules and looking at colocalization with the fluorescent probe lysotracker.

This work provides an avenue for the development of gene delivery systems based on GO-based nanoplatforms. Moreover, by taking advantage of the great qualities in terms of physicochemical, electrical and optical properties of GO, this study might provide novel strategies to overcome limitations commonly faced such as low stability of the translocating biomolecules and endosomal entrapment.

References:

[1] N. Nayerossadat, P. Ali, and T. Maedeh, "Viral and nonviral delivery systems for gene delivery," *Adv. Biomed. Res.*, vol. 1, no. 1, p. 27, 2012.

[2] D. Ibraheem, A. Elaissari, and H. Fessi, "Gene therapy and DNA delivery systems," *Int. J. Pharm.*, vol. 459, no. 1–2, pp. 70–83, 2014.

[3] M. M. Mady, "Cationic liposomes as gene delivery system," *African J. Pharm. Pharmacol.*, vol. 5, no. 17, pp. 2007–2012, Nov. 2011.

[4] J. L. Shirley, Y. P. de Jong, C. Terhorst, and R. W. Herzog, "Immune Responses to Viral Gene Therapy Vectors," *Molecular Therapy*, vol. 28, no. 3. Cell Press, pp. 709–722, 04-Mar-2020.

[5] M. Malmsten, "Inorganic nanomaterials as delivery systems for proteins, peptides, DNA, and siRNA," *Curr. Opin. Colloid Interface Sci.*, vol. 18, no. 5, pp. 468–480, 2013.

[6] N. D. Donahue, H. Acar, and S. Wilhelm, "Concepts of nanoparticle cellular uptake, intracellular trafficking, and kinetics in nanomedicine," *Adv. Drug Deliv. Rev.*, vol. 143, pp. 68–96, 2019.

[7] H. Hillaireau and P. Couvreur, "Nanocarriers' entry into the cell: Relevance to drug delivery," *Cellular and Molecular Life Sciences*, vol. 66, no. 17. Birkhauser Verlag Basel, pp. 2873–2896, 05-Jun-2009.

[8] E. Yuba *et al.*, "PH-sensitive polymer-liposome-based antigen delivery systems potentiated with interferon-γ gene lipoplex for efficient cancer immunotherapy," *Biomaterials*, vol. 67, pp. 214–224, Oct. 2015.

[9] A. Suarez-Arnedo *et al.*, "Novel BUF2-magnetite nanobioconjugates with cell-penetrating abilities," *Int. J. Nanomedicine*, vol. Volume 13, pp. 8087–8094, 2018.

[10] J. Sandoval *et al.*, *Carbon nanomaterials as pharmaceutic forms for sustained and controlled delivery systems*. 2019.

[11] M. N. Hasan, M. Nafiujjaman, and Y.-K. Lee, *2D Nanomaterials for Gene Delivery*. Elsevier Inc., 2019.

[12] N. Rahmanian, M. Eskandani, J. Barar, and Y. Omidi, "Recent trends in targeted therapy of cancer using graphene oxide-modified multifunctional nanomedicines," *Journal of Drug Targeting*, vol. 25, no. 3. Taylor and Francis Ltd, pp. 202–215, 16-Mar-2017.

[13] D. de Melo-Diogo, R. Lima-Sousa, C. G. Alves, E. C. Costa, R. O. Louro, and I. J. Correia, "Functionalization of graphene family nanomaterials for application in cancer therapy," *Colloids Surfaces B Biointerfaces*, vol. 171, no. May, pp. 260–275, 2018.

[14] P. Liu, S. Wang, X. Liu, J. Ding, and W. Zhou, "Platinated graphene oxide: A nanoplatform for efficient gene-chemo combination cancer therapy," *Eur. J. Pharm. Sci.*, vol. 121, no. March, pp. 319–329, 2018.

A.

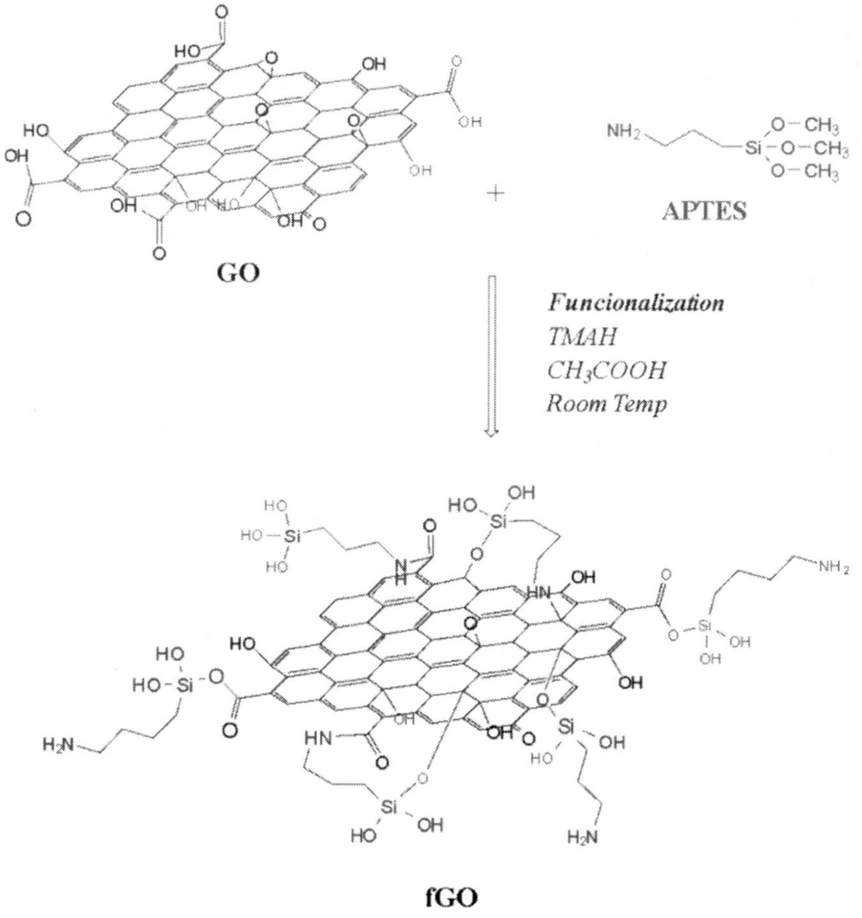

B.

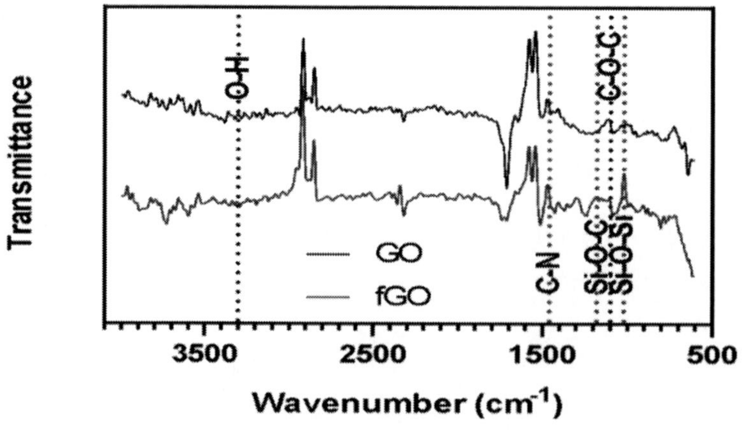

Fig. 1. Silanization of GO. (A) Scheme of the silanization reaction. This functionalization was conducted to render free amine groups on the surface of GO (B) FTIR analysis confirmed the presence of characteristic carboxyl groups on the surface of GO (black line) and the Si-O bond after silanization (blue line)

602230 Molecular Transport through Self-Assembled DNA Nanofluidic Channels

Abstract:

Nanoscale channels are a fundamental mechanism for directed transport within living systems. Confinement of transport to one dimension makes transport rapid, and gating at channel entrances can make transport selective. However, little is known about the mechanisms and rates of transport of small molecules within nanoscale channels, particularly those that extend beyond the dimensions of lipid membranes and across hundreds of nanometers or microns.

We have used techniques from structural DNA nanotechnology, which makes it possible to precisely control the geometry and chemical functionality of self-assembled structures, to construct synthetic nanoscale channels for molecular transport. The channel we have designed and constructed has an internal diameter of 4-6 nm and a length that can extend for multiple microns. It is composed of a DNA origami pore that penetrates and spans a lipid membrane and a DNA nanotube through which small molecules can transport. The nanotube self-assembles from DAE-E double crossover tiles, consisting of five DNA oligomers that grows from the DNA origami pore. Spontaneous lipid membrane incorporation is facilitated by hydrophobic functionalization of DNA origami pore via 12 cholesterol moieties. The transport through the channel can be stopped by adding a DNA origami cap that specifically binds to the channel opening and plugs the flow. These self-assembled channels can self-repair and can be grown such that their endpoints specifically attach at molecular landmarks, making them a promising biomolecular device for single molecule biosensing, for studying intercellular signaling, and for drug delivery.

We characterized molecular transport through both the DNA pore and DNA channels by confocal microscopy and a dye influx assay, in which synthesized DNA channels spontaneously inserted onto synthetic giant unilamellar vesicles. The rates of fluorescent dye transport through the channels were quantified by time-lapse observation of intensities inside many vesicles and comparing to a diffusion-dominated molecular transport model. Our results indicate that transport of small molecules through these channels is diffusive and that molecule size determines the degree to which transport may occur. We then designed a cap structure using the DNA origami technique, which has shown to bind to the open end of the channels at an over 95% yield. Experiments showed minimized molecular transport through capped DNA channels, suggesting that we have control over when the transport through the channels would occurs.

Figure 1. Scheme for studying the coherent transport of small molecules through a self-assembled DNA nanofluidic channel and the DNA nanostructure components used in the transport process. **a)** the goal of assembling the DNA nanofluidic channel is to study the mechanisms and rates of transport of small molecules within nanoscale channels. Whether the transport is coherent within the channel or the transport occurs across the channel boundary can be determined by characterizing the molecular transport rates when the channel is plugged with a DNA cap. **b)** to characterize the rates of molecular transport through a channel, the channel is inserted across a lipid membrane with a concentration gradient of small molecule fluorescent dyes across the membrane. The fluorescent dyes diffuse through the channel and enter the internal space of the membrane until reaching equilibrium. A cap that closes the channel opening would stop the transport if the dyes cannot transport across the channel boundary. **c)** a DNA origami nanostructure functionalized with hydrophobic moieties serves as a cross-membrane pore. DNA tiles hybridize to the DNA pore to self-assemble into a DNA nanotube that serves as the nanofluidic channel. A DNA origami nanostructure with minimal inner diameter specifically binds to the nanotube opening via complementary DNA 'sticky ends' to serve as a channel cap.

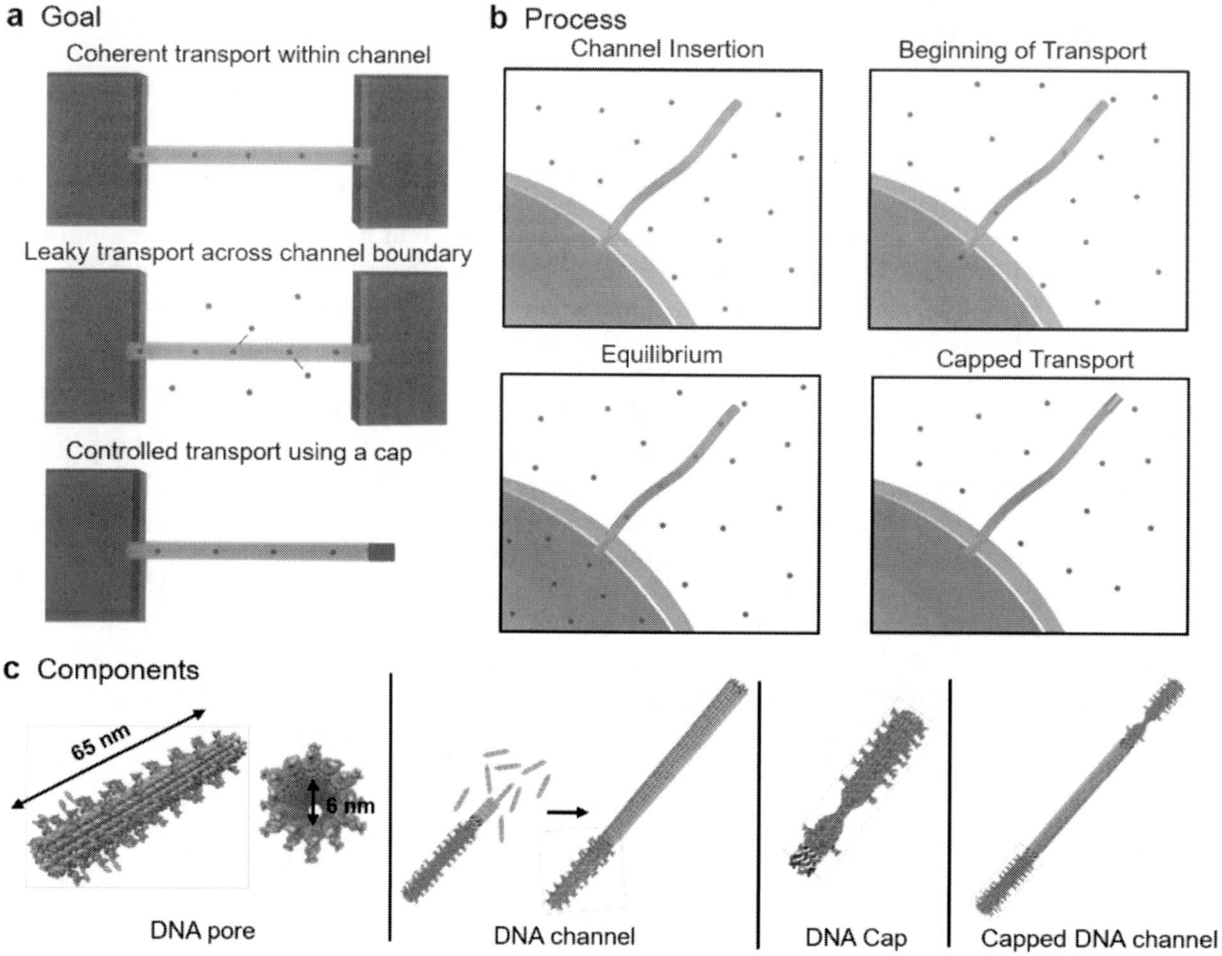

601502 Graphoepitaxy-Directed Crystallization of Small-Molecule Organic Semiconductors into 3D Structures

Abstract:

In this work, we apply the principle of graphoepitaxy to guide the solution-phase crystallization of model semiconducting perylene within nanopillar silicon grids. Graphoepitaxy-directed oriented growth of molecular crystals has previously been demonstrated through the matching of angles presented by surface relief structures to specific angles between crystallographic planes of molecular crystals. In contrast, our method relies on registry between the underlying grid periodicity and the overall shape of perylene crystals, which grow as flat plates. When nucleated within the nanopillar grids, perylene crystals adopted two distinct in-plane orientations in which the plates align either with the unconfined rows or columns between nanopillars. Ostwald ripening at the air/liquid/grid interface to dissolve smaller, misoriented crystals while enlarging crystals aligned with the grid structure resulted in the formation of three-dimensional trellises of single crystals meeting at 90^{0} angles. Building on our previous discovery of nanoporous scaffold-directed assembly of needlelike triisopropylsilylethynyl pyranthrene crystals into vertical crystal arrays[1], our findings point to the possibility of assembling single crystals into complex structures through judicious grid design for different crystal morphologies across materials classes. Critically, the high-energy p-planes of perylene crystals are exposed at the grid surface, which will have important implications for controlling material properties, from reactivity to dissolution rates.

[1] K. Zong,Y. Ma, K. Shayan, J. Ly, E. Renjilian, C. Hu, S. Strauf, A. Briseño, and S.S. Lee. "Directing Solution-Phase Nucleation to Form Organic Semiconductor Vertical Crystal Arrays." *Crystal Growth & Design* 19, (2019): 3461–3468.

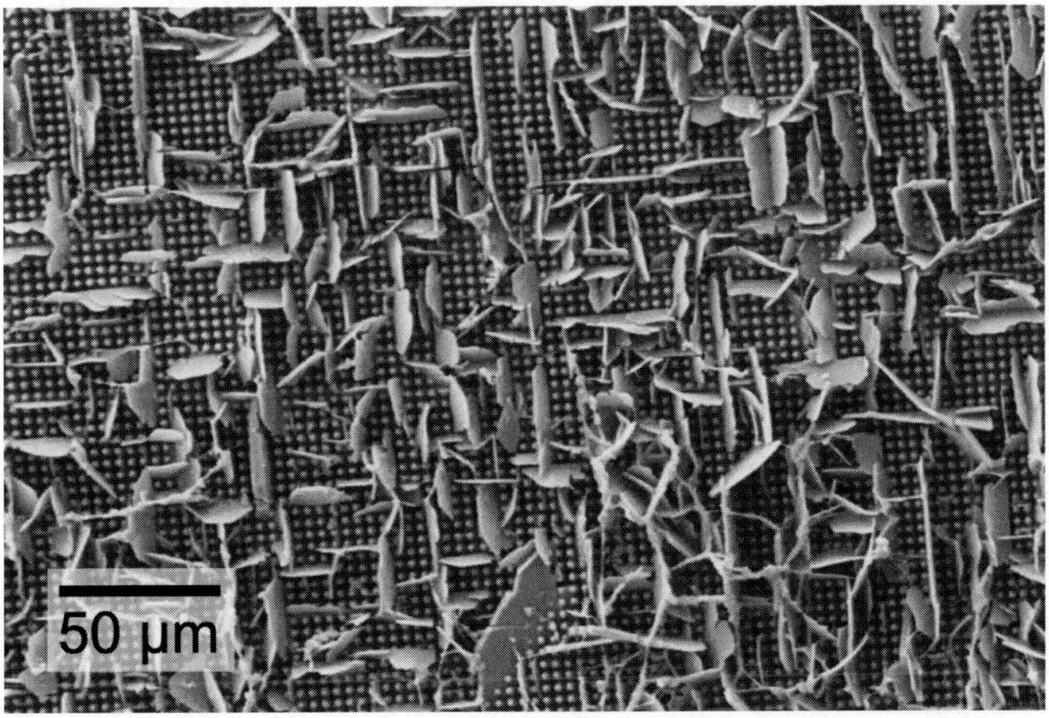

607478 Enhanced Heat Transfer on the Teflon Nanopore Surface Fabricated Using Anodized Alumina Templates.

Abstract:

Advanced energy systems rely on efficient cooling mechanisms for achieving higher power density and system efficiency. Liquid-vapor phase change heat transfer is one of the widely used cooling mechanisms in industries for thermal management. Improving the liquid-vapor phase change heat transfer efficiency could lead to substantial energy savings and reduced greenhouse emissions. Nanopatterning surfaces are known to enhance the liquid-vapor phase change heat transfer in the boiling and quenching regimes. The large surface area and the small features in nanopatterned surfaces allow more active sites for bubble nucleation and faster bubble departing frequency, thereby increasing the heat transfer efficiency. Nanopatterning polymers like Teflon are attractive due to their versatility and chemical resistance to most organic liquids and acids. In this work, we fabricated a Teflon nanopore array by a modified template-assisted pattern transfer technique using anodized aluminum oxide (AAO) as etch masks. The heat transfer performance of the fabricated Teflon nanopores was tested using droplet evaporation experiments with acetone as the working fluid. Traditional pattern transfer techniques using AAO on hydrophobic polymer substrates is limited to a smaller area of pore transfer. We identified the factors that hindered the scale-up process and followed a modified template transfer technique that can improve the pattern transfer on hydrophobic polymers like Teflon. The droplet evaporation experiments performed on the nanopore Teflon surface showed 2.5 times faster droplet evaporation rate than on a flat surface. The increased active sites for bubble nucleation, faster bubble departure frequency, and the increased maximum wetted area due to wet-Cassie type wetting of acetone is responsible for the enhanced heat transfer on the nanoporous Teflon.

608438 Graphene and MoS2 Gel for Aligned 3D Printing for Electronic and Mechanical Analysis

Abstract:

Designing 3D printed micro-architectures using electronic materials with well-designed electronic structures will potentially lead to accessible device fabrication for 'on-demand' applications. Here we show gelation of six different concentration of graphene/MoS2 (100:0, 10:90,30:70, 50:50, 70:30, 90:10) and Graphene/BN (50:50) alloys. Rheology studies show the viscoelastic properties of the gels, in addition to providing shield yield stress. The shear-thinning behavior provides the ability of fabrication of 3D printed devices in an aligned controlled nozzle-extrusion based modified 3D printer, enabling the fabrication of electrodes for analyzing electron-tunneling barrier width between conductive graphene-centers and studying the resultant mechanical properties. The 3D printed devices were solidified by removing water through lyophilization. The ordered graphitic region with sp^2 hybridized carbon atoms in the graphene sheets is in the order of 8.28 ±0.23 nm. The temperature-dependent electronic transport 3D printed electrodes structures exhibited a transport-barrier of 16 meV and a tunneling width of 0.56 nm (Fowler Nordheim electron tunneling) with graphene centers having a carrier concentration of $2.63457*10^{12}/cm^2$. By modifying the graphene/MoS2 concentration, we were able to control the Young's Modulus of the structure between 0.6 and 1.6 MPa. We envision that the proposed 3D-printing of gels of nanomaterial will lead to an evolution in the design of next-generation of 'on-demand' printed electronic and electromechanical devices.

604041 Dark-Field Microscopy for Sensing the Shape and Orientation of Isolated Gold Nanostructures

Abstract:

We demonstrate a robust and straightforward method to determine the shape and orientation of different gold nanostructures using dark-field microscopy (DFM). By simply analyzing the dark-field scattering images of the gold nanostructures with incidence polarized light in a bench-top DFM, the shapes and orientations of the underlying nanostructures can be accurately determined. Depending on the size, shape, and orientation of plasmonic nanoparticles, which dictate the plasmonic behaviors on a dielectric substrate, the nanoparticles appear as a distribution of color as captured by a commercial, color CCD camera. Extracting the red, green, and blue channels of the scattering signal from a CCD imaging detector as a function of azimuthal angle provides a polar plot of color intensity values that is characteristic of the shape of the underlying nanostructure. Using this technique, we have been able to sense the shape and orientation of different gold nanostructures, including nanorods and nanowires, spherical nanoparticles, nanoparticle dimers, and nanotriangles. We demonstrate that the polarization-dependent scattering can be used to quickly identify the shape of individual objects as well as the orientations of vastly populated objects on a substrate. We anticipate that this technique will provide a rapid and inexpensive complement to the typical structural analysis of nanoparticles that is achieved by electron microscopy.

609666 Atomic Ordering in a Monolayer TMD Alloy

Abstract:

Distribution of atomic species in an alloy can be a determining factor in the chemical properties of the material. Advances in the fabrication and imaging of two-dimensional materials, in particular, transition-metal dichalcogenides (TMDs), have allowed researchers to successfully produce TMD alloys and precisely observe the atomic species distribution using scanning transmission electron microscopy (STEM). When the atomic number of the two transition-metals are sufficiently different, they can be distinguished clearly in high-resolution STEM images. Here, we report the observation of short-range atomic ordering in a NbS_2-ReS_2 alloy, and describe the properties of this ordering. [1] It turns out that the origin of the short-range ordering is geometric frustration, which occurs when the constituent interactions of a system cannot be simultaneously satisfied due to the underlying geometry. We find that this system is analogous to the 2D Ising model of antiferromagnetic spins in a triangular lattice, and long-range order is thoroughly suppressed. Using density functional theory, we build a lattice model that closely matches the statistical properties of the experimental atomic distribution. We also demonstrate the effects of atomic ordering on the band structure. We predict that the degree of ordering can be controlled by the growth temperature, allowing the tuning of the chemical properties of the material. This alloy has excellent ambient stability, a highly thickness-dependent bandgap, and promising NO_2 sensing properties. [2]

1. Azizi, **M. Dogan**, J. D. Cain, R. Eskandari, X. Yu, E. C. Glazer, M. L. Cohen, A. Zettl, "Frustration and Atomic Ordering in a Monolayer Semiconductor Alloy", *Phys. Rev. Lett.* **124**, 096101 (2020).
2. Azizi, **M. Dogan**, H. Long, J. D. Cain, K. Lee, R. Eskandari, A. Varieschi, E. C. Glazer, M. L. Cohen, A. Zettl, "High-Performance Atomically-Thin Room-Temperature NO_2 Sensor", *arXiv*:2003.01909 (2020).

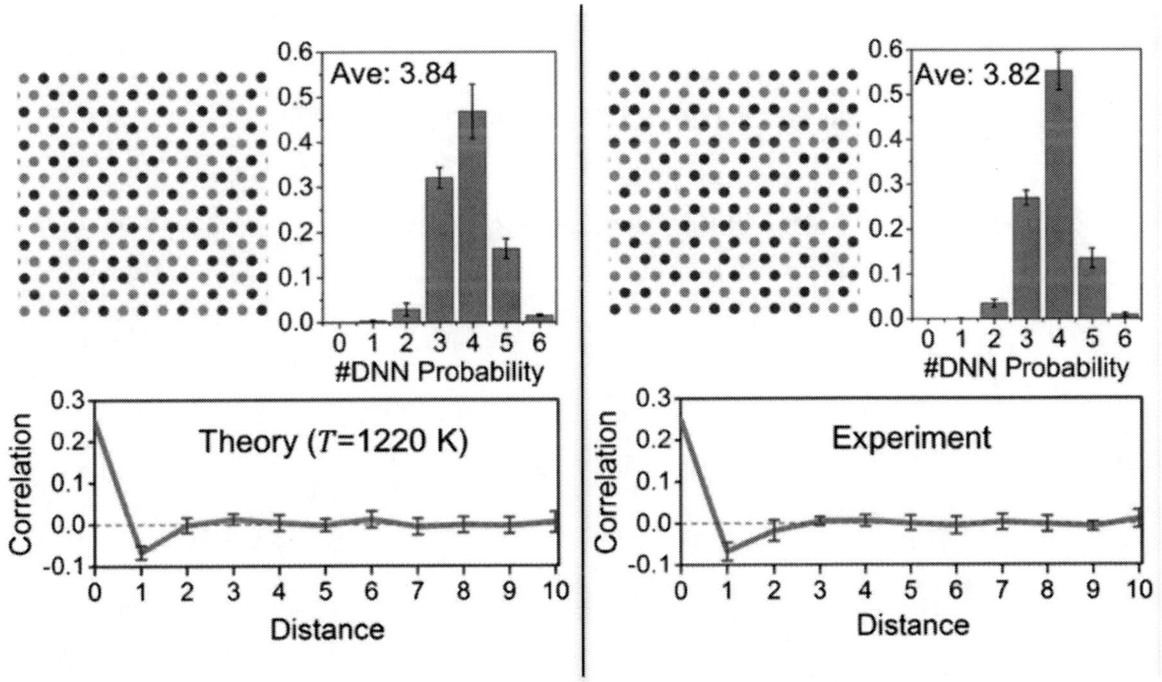

611159 Visualizing Oxidation Mechanisms in Few-Layered Black Phosphorus Via *in Situ* Transmission Electron Microscopy

Abstract:

Layered two-dimensional (2D) black phosphorus (BP) exhibits novel semiconducting properties including a tunable bandgap and high electron mobility. However, the poor stability of BP in ambient environment severely limits potential for application in future electronic and optoelectronic devices. While passivation or encapsulation of BP using inert materials/polymers has emerged as a plausible solution, a detailed fundamental understanding of BP's reaction with oxygen is imperative to rationally advance its use in applications. Here, we use *in situ* environmental transmission electron microscopy to elucidate atomistic structural changes in mechanically exfoliated few-layered BP during exposure to varying partial pressures of oxygen. An amorphous oxide layer is seen on the actively etching BP edges, and the thickness of this layer increases with increasing oxygen partial pressure, indicating that oxidation proceeds via initial formation of amorphous P_xO_y species which sublime to result in the etching of the BP crystal. We observe that while few-layered BP is stable under the 80 kV electron beam (e-beam) in vacuum, the lattice oxidizes and degrades at room temperature in the presence of oxygen only in the region under the e-beam. The oxidative etch rate also increases with increasing e-beam dosage, suggesting the presence of an energy barrier for the oxidation reaction. Preferential oxidative etching along the [0 0 1] and [0 0 1] crystallographic directions is observed, in good agreement with density functional theory calculations showing favorable thermodynamic stability of the oxidized BP (0 0 1) planes compared to the (1 0 0) planes. We expect the atomistic insights and fundamental understanding obtained here to aid in the development of novel approaches to integrate BP in future applications.

References:

Naclerio et al. ACS Appl. Mater. Interfaces 2020, 12, 13, 15844-15854, DOI:10.1021/acsami.9b21116

601373 Scaffolded DNA-Dye Complexes: Theory of Molecular Interactions for Synthetic Light Harvesting Applications

Abstract:

Natural photosynthetic systems utilize quantum coherent effects to enhance energy transfer efficiency when light-harvesting molecules are closely interacting. These natural photosynthetic complexes enable molecular packing and interaction by precise scaffolding within a protein environment. Likewise, synthetic DNA nanotechnology enables control of molecular interactions by rational attachment of photoactive dyes to DNA scaffolding. In both natural and synthetic systems, closely-interacting dyes may exhibit J- (red-shifted energies) and H-aggregation (blue-shifted energies) behavior, effects which are caused by excitonic delocalization of the molecular wavefunctions. Using model systems consisting of cyanine dyes attached or interacting with DNA scaffolding, we explore a range of excitonic and structural behavior using a combined quantum mechanics (QM) and molecular dynamics (MD) theoretical toolset. We explore complexes with cyanine dyes non-covalently bound to DNA, as well as complexes with cyanine dyes covalently attached to the DNA backbone. Using MD, we explore the dynamical motion of these cyanine-DNA complexes, and subsequently using QM, the visible absorption spectra and photoactive properties upon aggregation can be calculated and compared to experimental results. Thus, using theoretical tools, we can build a molecular exciton model to elucidate the environmental effect of DNA on aggregation and excitonic behavior of photoactive dyes. These excitonic and structural aggregation properties of cyanine-based DNA-dye complexes can be compared to natural light-harvesting complexes with protein-based scaffolding. Understanding the differences between natural and synthetic biological scaffolds are of major interest, and knowledge of excitonic behavior at the molecular level will enable directed bio-inspired design of scaffolded photoactive materials.

599215 Self-Organized Protein Arrays Guided By DNA Origami Lattices

Abstract:

DNA nanotechnology has offered an exciting direction for bio- and nano-material developments, by creating bottom-up frameworks to program and construct macromolecular objects with nano-scale resolution and spatial addressability. In this work, we report a design of functional 2D and 3D protein arrays using the structure programmable DNA origami framework. We selected ferritin, an iron storage protein, and demonstrated a two-step covalent conjugation and DNA hybridization for high encapsulation efficiency of ferritin inside the 3D wireframed octahedral DNA origamis (Octa). The position of encapsulated proteins was tailored inside the structure while preserving the conjugates and the overall Octa structure. For further supramolecular assembly, proteins were encapsulated in uniquely designed vertex-encoded Octa and assembled by the vertex-to-vertex interconnection to create 2D single-layered, double-layered and 3D lattices with controllable and tailorable lattice structures and parameters. Furthermore, stability and functionality of the protein arrays built in the precisely designed 3D DNA scaffold were accessible for small molecule transport to remove the iron core of ferritin, while maintaining the lattice structure during the conversion of ferritin into its core-less form, apoferritin. The work provides a comprehensive and detailed analysis of the high-order protein/origami lattice structures by *in situ* small angle X-ray scattering, electron microscopy and computer modelling, to reconstruct the lattices in 3D and extract structural and conformational details.

610917 Microgliosis in a Dish

Abstract:

Microgliosis, due to brain injury or disease, can lead to varying molecular and morphological changes in microglia. Magnetic resonance elastography and ultrasound studies show that brain stiffness dynamically changes with age and disease state. Molecular mechanisms that account for the stiffness predilection to brain dysfunction, specifically on microglia, during brain disorders have been underexplored, and the impact of stiffness on microglia dysfunction and alterations during brain trauma is a gap in knowledge. The central goal of this study is to understand the role of mechanotransduction (stiffness) in microglia molecular and functional remodeling leading to increase brain dysfunction risk.

In this study, we have developed an innovative biomimetic platform named "BEASTS (Bio-Engineered Adhesive Siloxane substrate with Tunable Stiffness)" based on polydimethylsiloxane (PDMS) substrate in combination with our patented polyelectrolyte multilayer film (PEM)-coating technology to engineer mechanically tunable substrates to mimic physiologic (2 kPa),; early microgliosis (8 kPa); and diseased (25 kPa) stiffness to investigate the effect of stiffness on microglia activation and function. We observed that microglia grown on soft substrates (2 kPa) displayed a consistently more quiescent phenotype while those on stiff substrates (8 kPa and 25 kPa) displayed microgliosis-like morphology. In addition to morphological changes, microglia cultured on thick substrates demonstrated a significant increase in other microgliosis hallmarks- cellular proliferation and growth. Furthermore, culturing microglia on stiff surfaces resulted in increased reactive oxygen species (ROS) production. A similar effect has been observed in microglia in brain trauma animal models and patients. These data suggest a plausible mechanism that increased stiffness modulates microglia function, causing brain injury. Our platform lends itself to the study of potential therapeutic strategies for brain injury focusing on the intricate brain microenvironment-microglia signaling pathways.

609490 *In Situ* Generation of Metal-Oxide Nanoparticles on Top of a Green-Synthesized Tellurium Nanowire Template and the Biomedical Study of the Synergetic Structure

Abstract:

Two of the major concerns that the healthcare system is facing nowadays are cancer and antimicrobial resistance (AMR) to antibiotics. Nanotechnology appears as a suitable solution, which might overcome some limitations of current available treatments. Despite of the advances in the nanoscale, there is a need to find alternatives to the traditional synthesis of nanomaterials, which suppose a threat to both the environment and society. In this context, Green Nanotechnology is presented as an answer, with cost-effective and environmentally-friendly approaches for nanoparticles synthesis.

In the present work, starch-mediated Tellurium nanowires were employed as a template for the *in-situ* growth of palladium and platinum nanostructures. The noble metal-chalcogenide nanocomposites were characterized for their biomedical applications, with both green-mediated synergetic composites showing antibacterial activity against AMR bacterial strains, both Gram negative (MDR *Escherichia coli*) and positive (Methicillin resistant *Staphylococcus aureus*) bacteria, at concentrations from 10 to 100 μg/mL over a 24-hour time period. Moreover, cell studies were done with human dermal fibroblast (HDF) and melanoma cells for 5 days, showing no significant cytotoxic effect at concentrations up to 25 μg/mL, while triggering a dose-dependent anticancer effect in the same rage of concentrations. Therefore, the use of noble metal-chalcogenide nanocomposites is proposed as a novel green nanotechnological-based platform for biomedical applications.

607096 Functionalized Mesoporous Silicas Direct Structural Polymorphism of Amyloid-β Fibrils

Abstract:

The aggregation of Amyloid-β (Aβ) is associated with the onset of Alzheimer's Disease (AD) and involves a complex kinetic pathway as monomers self-assemble into fibrils. A central feature of amyloid fibrils is the existence of multiple structural polymorphs, which complicates the development of disease-relevant structure-function relationships. Developing these relationships requires new methods to control fibril structure. In this work, we demonstrate that mesoporous silicas (SBA-15) functionalized with hydrophobic (SBA-PFDTS) and hydrophilic groups (SBA-PEG) direct the aggregation kinetics and resulting structure of $A\beta_{1-40}$ fibrils. The hydrophilic SBA-PEG had little effect on amyloid kinetics while as-synthesized and hydrophobic SBA-PFDTS accelerated aggregation kinetics. Subsequently, we quantified the relative population of fibril structures formed in the presence of each material using electron microscopy. Fibrils formed from $A\beta_{1-40}$ exposed to SBA-PEG were structurally similar to control fibrils. In contrast, $A\beta_{1-40}$ incubated with SBA-15 or SBA-PFDTS formed fibrils with shorter cross-over distances that were more structurally representative of fibrils found in AD patient-derived samples. Overall, these results suggest that mesoporous silicas and other exogenous materials are promising scaffolds for the *de novo* production of specific fibril polymorphs of $A\beta_{1-40}$ and other amyloidogenic proteins.

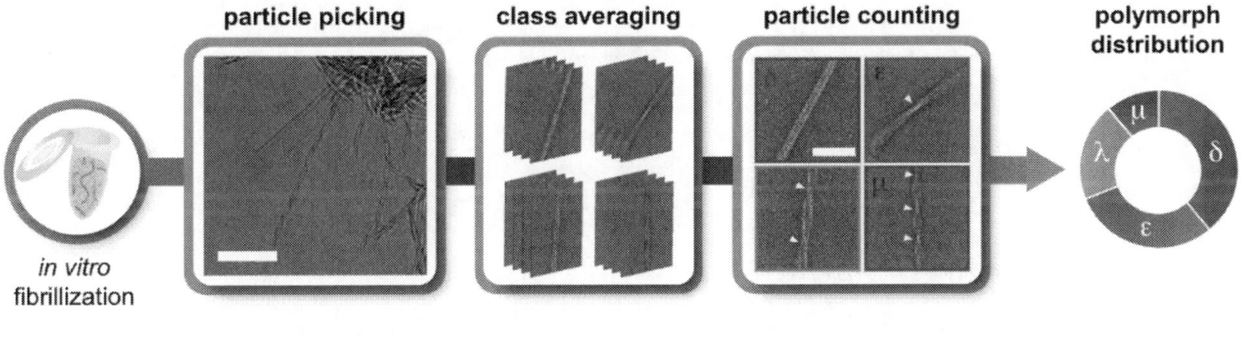

597983 Investigation of the Antifouling Properties of Polyproline Self-Assembled Monolayers

Abstract:

Biofouling is a major concern for the surfaces of medical devices and biosensors. The goal of this research was to quantify protein adsorption onto polyproline to determine its capabilities in preventing biofouling on those surfaces. A quartz crystal microbalance with dissipation (QCM-D) was used to measure the protein adsorption onto a gold sensor functionalized with a polyproline self-assembled monolayer. The adsorption of bovine serum albumin (BSA) and human plasma fibrinogen onto the coating was measured by examining the change in frequency shift, which is directly proportional to the change in mass. The measured change in frequency due to BSA adsorption was approximately ten times less on a polyproline SAM than on bare gold. The change in frequency due to fibrinogen adsorption was reduced from 100-125 Hz on bare gold to 40-45 Hz on the peptide coating. The data indicate that polyproline has the potential to repel proteins and therefore reduce biofouling. Moreover, the use of polyproline demonstrates that peptides can be used to further examine properties of antifouling surfaces, as different modifications of this peptide can be tested using this methodology.

597879 Determining SWCNT Extraction Conditions in Aqueous Two-Polymer Phase Extraction with Near-Infrared Fluorescence

Abstract:

Aqueous two-polymer phase extraction (ATPE) is an effective, yet still developing, separation method for enriching or isolating single species of single-wall carbon nanotubes (SWCNTs). However, methods for rapidly determining the aqueous surfactant concentration conditions that yield upper or lower polymer phase extraction for specific (n,m) species SWCNTs in ATPE separations are a missing component limiting quicker development and optimization of the ATPE technique. Here, we describe the use of near-infrared fluorescence measurements on single-phase dispersions of SWCNT species mixtures to determine these partition coefficient conditions without performing ATPE separations. This new methodology enables dramatically swifter determination of partitioning conditions for individual (n,m) species and their handed enantiomers, while using dramatically less material and without the ambiguity of mass transfer limitations or interfacial absorption present when conducting the ATPE separation itself. The ability to fully describe the partitioning conditions for each (n,m) and (m,n) SWCNT species is expected for multiple surfactant composition competitions and surfactant variants. Investigation of complicating factors such as temperature effects, polymer compositions, SWCNT properties, and the redox environment are likewise enabled by this methodology and will also be described.

609251 Improvements in Aqueous Two-Polymer Phase Extraction of Single-Wall Carbon Nanotubes Beyond the Small-Diameter Regime

Abstract:

Extending aqueous two-phase extraction methods for effective separation of single-wall carbon nanotubes outside the small-diameter regime is a challenging effort due to the increasing number of species and lesser mass fractions of each component in synthetic mixtures. In this talk I will detail the use of several new developments, including the use of co-surfactant mixtures, and the prior analytical determination of partition coefficients for particular (n,m) species, to advance separation efficiency in this regime. Examples demonstrating designed isolation of specific (n,m) structures to the level of single enantiomers will be presented, as well as the characterization enabling such experimental design.

599430 Molecular-Dynamics Analysis of the Mechanical Behavior of Graphene Nanomeshes

Abstract:

Graphene nanomeshes (GNMs), also known as nanoporous graphene, are ordered defect-engineered graphene nanostructures consisting of periodic arrangements of nanopores in the graphene lattice with neck widths between pores of less than 10 nm, typically fabricated by block copolymer lithography and plasma/electron etching of graphene. The unique porous structure of GNMs and their outstanding mechanical and electronic properties makes these 2D materials very appealing for numerous technological applications, ranging from field effect transistors to membranes used in water desalination. Here, we present a systematic study of the response of GNMs to mechanical loading based on molecular-dynamics (MD) simulations according to a reliable bond order interatomic potential and establish the corresponding structure-properties relationships.

We report results on the response of graphene nanomeshes (GNMs) to uniaxial tensile straining based on MD simulations of dynamic deformation tests. We examine the effects on the GNM mechanical behavior under straining along different directions of the nanomesh pore morphology and pore edge passivation by testing GNMs with elliptical pores of various aspect ratios and different extents of edge passivation through termination with H atoms of under-coordinated edge C atoms. We establish the dependences of the ultimate tensile strength, fracture strain, and toughness of the GNMs on the nanomesh porosity, derive scaling laws for GNM strength-density relations, and find the GNMs' mechanical response to uniaxial straining to be anisotropic for pore morphologies deviating from circular. We also find that the GNM tensile strength decays exponentially with increasing GNM porosity and that pore edge termination with H atoms causes a reduction in the GNMs' elastic stiffening, strength, deformability, and toughness; this hydrogen embrittlement effect is more pronounced at a high level of pore edge passivation that renders the edge C atoms sp^3-hybridized.

We also report results on the mechanical and structural response of GNMs to indentation based on MD simulations of nanoindentation tests. We find the GNMs' response to indentation to be nonlinearly elastic until fracture initiation, with elastic properties that depend strongly on the GNM porosity but are not sensitive to pore edge passivation, which, however, influences the GNM failure mechanism past fracture initiation. Increasing GNM porosity leads to a monotonic decrease of the 2D elastic modulus of the GNMs and the modulus-porosity dependence follows a quadratic scaling law. The maximum stress reached at the GNM breaking point is high throughout the porosity range examined, even at very high porosity. The maximum deflection of the indented GNMs at their breaking point exhibits a minimum at porosities below 20%; beyond this critical porosity, the maximum deflection increases monotonically with increasing porosity and can reach values comparable to half of the indented sample radius at high porosities. Moreover, our analysis reveals a novel inelastic, dissipative necking mechanism of GNM failure at high porosities that further enhances the excellent deformability of the GNMs. Our findings highlight the potential of graphene nanomeshes as 2D mechanical metamaterials whose mechanical response can be tuned by proper tailoring of their structural features.

611135 Analytic Approach to Magneto-Strain Tuning of Electronic Transport through a Graphene Nanobubble: Perspectives for a Strain Sensor

Abstract:

We consider the scattering of Dirac particles in graphene due to the superposition of an external magnetic field and mechanical strain. As a model for a graphene nanobubble, we find exact analytical solutions for single-particle states inside and outside a circular region submitted to the fields. Finally, we obtain analytical expressions for the scattering cross-section, as well as for the Landauer current through the circular region. Our results provide a fully-analytical treatment for electronic transport through a graphene nanobubble, showing that a combination of a physical magnetic field and strain leads to valley polarization and filtering of the electronic current. Moreover, our analytical model provides an explicit metrology principle to measure strain by performing conductance experiments under a controlled magnetic field imposed over the sample.

References

E. Muñoz and R. Soto-Garrido, Journal of Physics: Condensed Matter 29 (2017) 445302

598587 High-Yield Analysis of Individual Ions and Molecules through the Interior of Carbon Nanotubes

Abstract:

The interior channels of carbon nanotubes are promising for studying transport of individual molecules in a 1D confined space. However, experimental investigations of the interior transport have been limited by the extremely low yields of fabricated nanochannels and their characterization. Here, this challenge is addressed by assembling nanotube membranes on glass capillaries and employing a voltage-ramping protocol. Centimeter-long carbon nanotubes embedded in an epoxy matrix are sliced to hundreds of 10 μm-thick membranes containing essentially identical nanotubes. The membrane is attached to glass capillaries and dipped into analyte solution. Repeated ramping of the transmembrane voltage gradually increases ion conductance and activates the nanotube ion channels in 90% of the membranes; 33% of the activated membranes exhibit stochastic pore-blocking events caused by cation translocation through the interiors of the nanotubes. Since the membrane-capillary assembly can be handled independently of the analyte solution, fluidic exchange can be carried out simply by dipping the capillary into a solution of another analyte. This capability is demonstrated by sequentially measuring the threshold transmembrane voltages and ion mobilities for K^+, Na^+, and Li^+. Beyond the detection of solvated ions, varying the diameter of carbon nanotubes enables analysis of analytes with different sizes. We show, for instance, detection and identification of individual polymeric molecules with different chain lengths by using a carbon nanotube with diameter of ~3 nm. This approach, validated with carbon nanotubes, will save significant time and effort when preparing and testing a broad range of solid-state nanopores.

610289 Study of Unique Thermodynamic Equilibria of Co-Surfactant Systems for Functionalizing Specific Chiralities of Single Walled Carbon Nanotubes

Abstract:

Single walled carbon nanotubes (SWCNTs) find applications in electronics and biosensors owing to their unique optical properties. Current biomedical applications are restricted due to constraints in separating highly monochiral fractions. Surfactant structure around the nanotube plays an important role in stability and separation of (n,m) species. The presence of thermodynamically stable co-surfactants states enables selective coating of species in solution. Our group has previously utilized this approach to elute (6,5) species at at very specific critical exchange ratio concentrations (CERC). The co-surfactant solution utilizes a ratio of sodium dodecyl sulphate (SDS)/sodium deoxycholate (DOC) to obtain single-chiral eluents of (6,5) species. This idea was further extended to obtain CERC values of different (n,m) species for the SDS-DOC co-surfactant system.

Here, we present utilization of this thermodynamic equilibria for selectively eluting a mixture of chiralities and modifying their functionality. Adsorption of added surfactant leads to reorganization of the surfactant shell around nanotube which leads to solvatochromatic shifts that can be probed through optical spectroscopy. Addition of DOC to SWCNT solution results in elution of a mixture of chiralities whose CERC is below the SDS/DOC ratio. Addition of SDS to this mixture results in opposite solvatochromatic shifts whose extent would indicate the surfactant reorganization around a particular (n,m) species. This would enable to functionalize selective species in solution. Different ligands were studied for functionalizing different chiralities for simultaneous detection of analytes in solution. This approach provides a promising pathway for using SWCNTs as biosensors by selectively functionalizing different chiralities without obtaining monochiral fractions.

605587 Laser Induced Graphene on Polysulfone Membrane for H_2/CO_2 Gas Separation

Abstract:

There has been a growing interest in polymeric membranes for gas separation due to their low cost, low energy usage, and lack of moving parts. Dense polymeric membranes have an intrinsic trade-off between selectivity and permeability based on the kinetic diameter of the gases involved. Molecular sieving materials like graphene-based nanomaterials, MXenes, zeolites, and carbon nanotubes. shown promising applications in gas separation membranes. Mixed matrix membranes with improved selectivity over polymeric membranes have been developed by adding nanomaterials as fillers in polymer matrix. However, the membrane fabrication is limited by dispersion of fillers because of their incompatibility with polymer matrix and the membrane performance is compromised due to unselective voids. [1] Laser induced graphene (LIG) can be photothermally formed on surface of polymeric substrates; this method is fast and scalable and has been applied for ultrafiltration applications in polysulfone, polyethersulfone, and polyphenylsulfone.[2]

Here, we report that LIG on surface of polysulfone substrates can enhance the H_2/CO_2 perm-selectivity of the control polysulfone film. A new class of heterogeneous mixed matrix membranes is developed by forming graphene on surface of polymer and combining molecular sieving mechanism with solution-diffusion mechanism of baseline polymers. In a proof-of-concept experiment, we have used LIG on 100 micrometers thick polysulfone substrate for separation of light gas pair H_2/CO_2 at room temperature and 1 atm pressure. Preliminary measurements indicate that this combination is a high perm-selectivity membrane with a selectivity of 51 and permeance of 707 GPU (1 GPU = 3.35 x 10^{-10} mol/m^2/Pa) as compared to neat polysulfone with selectivity of 2 and permeance of 1 GPU. X-Ray Diffraction analysis, thermogravimetric data, optical and scanning electron microscopy were utilized to study the final material; these characterizations confirmed graphene formation on the substrate.

References:

[1] M. Etxeberria-Benavides, O. David, T. Johnson, M. M. Łozińska, A. Orsi, P. A. Wright, S. Mastel, R. Hillenbrand, F. Kapteijn, J. Gascon, *Journal of Membrane Science* **2018**, 550, 198.

[2] S. P. Singh, Y. Li, J. Zhang, J. M. Tour, C. J. Arnusch, *ACS Nano* **2018**, 12, 289.

614025 (Invited Talk) Counting Molecules, Dodging Blood Cells: Continuous, Real-Time Molecular Measurements Directly in the Living Body

Abstract:

The availability of technologies capable of tracking the levels of drugs, metabolites, and biomarkers in real time in the living body would revolutionize our understanding of health and our ability to detect and treat disease. Imagine, for example, a dosing regime that, rather than relying on your watch ("take two pills twice a day"), is instead guided by second-to-second measurements of plasma drug levels wirelessly communicated to your smartphone. Such a technology would likewise provide researchers and clinicians an unprecedented window into neurology and physiology, and could even support ultra-high-precision personalized medicine in which drug dosing is optimized minute-by-minute using closed-loop feedback control. Towards this goal, we have developed a biomimetic,electrochemical sensing platform that supports the high frequency, real-time measurement of specific molecules (irrespective of their chemical reactivity) in situ in the blood and tissues of awake, freely moving subjects. Key to achieving this was the development of sensors small enough to emplace in the body but retaining sufficient s/n in such a complex environment. This was done via the use of nanostructured and nanoporous electrode materials of greatly enhanced microscopic area.

610717 Ultrasensitive, Selective, and Reversible Room-Temperature NO_2 Sensor Based on a Monolayer Transition Metal Dichalcogenide

Abstract:

Chemical sensors are essential for environmental and health monitoring. In particular, gas sensors that can efficiently detect toxic gases emitted from power plants, vehicles, and industrial sources at room temperature, are of substantial importance. Here, we demonstrate a room-temperature NO_2 sensor based on a monolayer transition metal dichalcogenide that is highly sensitive and selective. Depending on the layer number of the sensing devices, their electrical resistances substantially increase or decrease upon exposure to NO_2. We find the atomically-thin sensors to be strongly selective to NO_2 with only minimal responses to NH_3, CH_2O, and CO_2. Importantly, the monolayer sensor shows complete reversibility with fast recovery in ambient condition (*i.e.* at room temperature and under substantial relative humidity). Combined with theoretical calculations, we explore interactions between different molecular species and the monolayer crystal and discuss the underlying sensing mechanism.

605192 Atomically-Thin Sensing Surfaces from 2D Materials for Detecting Cellular Gaps

Abstract:

Cellular layers play critical role as barriers, filters, and membranes in our bodies. The performance of cellular layers is often found to be determined by the properties of individual cellular gaps. Yet these gaps often fall below the diffraction limit, hampering their detection. Here, we develop a novel sensing modality called chemical tomography, where the diffusion of a harmless chemical tracer is used to reveal cellular gaps in human umbilical vein endothelial cellular (HUVEC) layers. For the readout, we employ atomically-thin and optically-active MoS2, whose two-dimensional form-factor enables it to serve as an array of 20,164 wireless sensors over 420x420 μm^2 area. A rigorous screening of 30 analytes identifies several tracer candidates with MoS2 dynamic range over five orders of magnitude in tracer concentration. In a single layer of HUVEC cells, we find local spots with 10^{-8}-10^{-9} cm^2/s diffusivities that are spatially correlated with immunofluorescence of vascular endothelial (VE)-cadherin proteins found on cell membrane. These spots correspond to cellular pores with effective sizes from 50 to 100 nm. This platform enables future studies on properties and dynamics of cellular gaps of various tissues.

607027 Nanocomposites of Multiwalled Carbon Nanotubes and Palladium-Based Nanosheets for Hydrogen Sensing

Abstract:

Thin palladium nanosheets have been gaining attention for various applications due to their high surface area, high surface atom fraction, and excellent electrical conductivity. Palladium has been explored as a resistive hydrogen sensor because of its unique property of forming hydrides in the presence of hydrogen at room temperature. However, most Pd-based sensors suffer from low response with a limit of detection around 300 ppm. This work describes the design and fabrication of multi-walled carbon nanotubes@Pd nanosheets (MWCNT@Pd NS) composites for hydrogen sensing applications. The Pd nanosheets were prepared by a CO confinement strategy using a one-pot solvothermal method. The prepared Pd nanosheets had a high surface area of ~125 m^2/g with a thickness of only ~2 nm. The prepared Pd nanosheets were then dispersed in hexane along with CNT and sonicated to produce a homogeneous suspension. The nanocomposite solution was drop cast onto an interdigitated electrode, and its performance for hydrogen detection in synthetic air was evaluated. Compared with pristine Pd nanosheets, the MWCNT@Pd nanosheets showed enhanced hydrogen detection performance. The resistance change of MWCNT@Pd nanosheets to 2% H_2 was 2%, compared to 0.5% change for pristine Pd nanosheets. The LOD of MWCNT@Pd nanosheets was also found to improve to <10 ppm, far below the 300 ppm LOD of Pd nanosheets. This encouraging result could prove useful in the fabrication of low-cost hydrogen sensors based on 2D Pd nanomaterials.

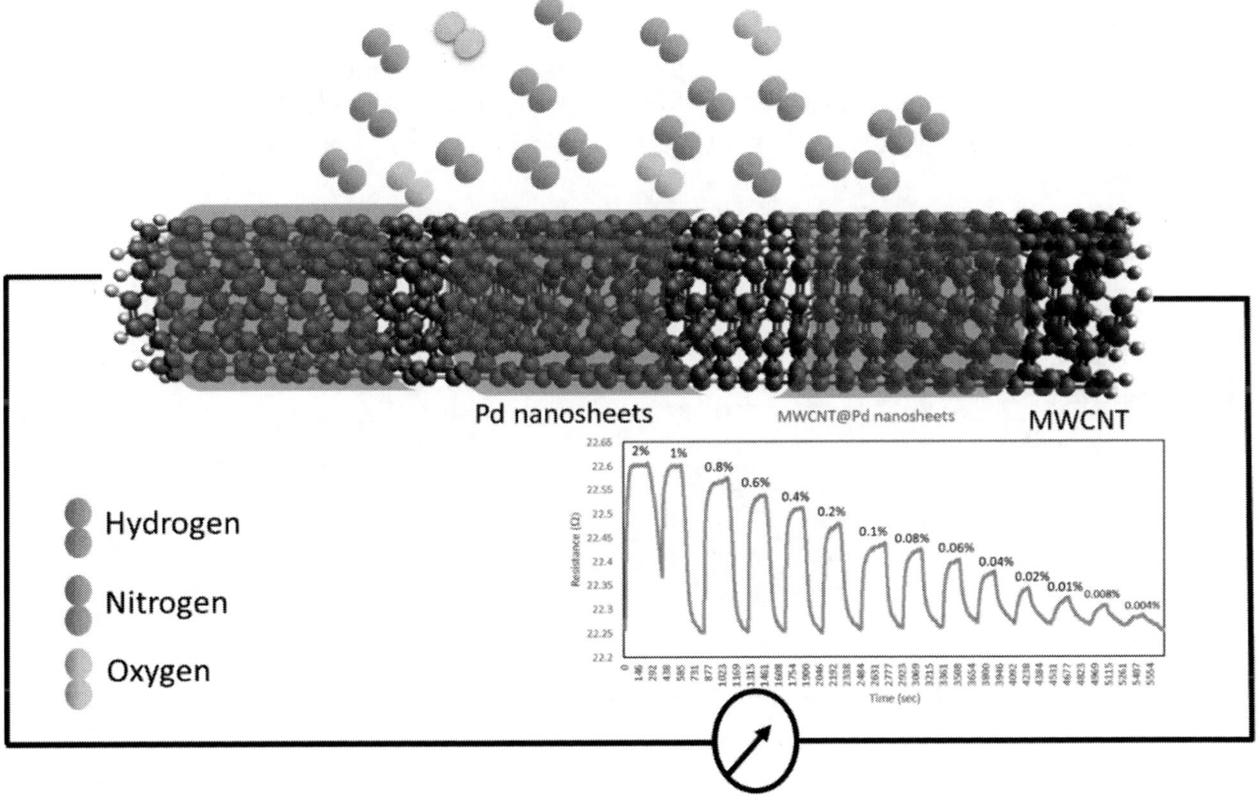

608092 Flame Aerosol Synthesis of Palladium-Decorated Crumpled Reduced Graphene Oxide Nanocomposites for Hydrogen Detection at Room Temperature

Abstract:

Hydrogen (H_2), a green energy source, is highly flammable with a lower explosive limit of just 4% in air. Its leakage during production, transportation, and storage can result in disastrous consequences. Therefore, development of accurate, rapid, and stable H_2 sensors that work in air at room temperature is crucial. In this presentation, we will describe a novel three-dimensional palladium (Pd)-decorated crumpled reduced graphene oxide ball (Pd-CGB) nanocomposite for H_2 detection in air at room temperature. Pd-CGB nanocomposites were synthesized using a flame-driven high temperature reducing jet (HTRJ) process. In this process, a hydrogen-rich flame passes through a converging-diverging nozzle. An aqueous precursor solution containing graphene oxide and palladium nitrate is injected at the throat section of the nozzle and is atomized by the high velocity gas stream. The resulting droplets evaporate, graphene oxide sheets crumple, and the palladium precursor decomposes, initiating nucleation of palladium nanoparticles on crumpled graphene oxide in a reducing environment containing excess H_2. After the reaction zone, nanocomposites are cooled immediately to prevent further growth and coalescence of the Pd nanoparticles. Hydrogen sensors produced by simple drop-casting of these nanocomposites onto electrodes performed well over a wide range of H_2 concentrations (0.01-2%) with response (resistance change) value, response time, and recovery time of 14.8%, 73 s, and 126 s, respectively, at 2% H_2. Compared with other reports on Pd-graphene sensors, the present study shows superior response/recovery time. The HTRJ process enables the single-step, continuous production of Pd- and other metal-decorated graphene nanocomposites with great potential for creating various gas sensors by simple drop-casting onto low-cost electrodes.

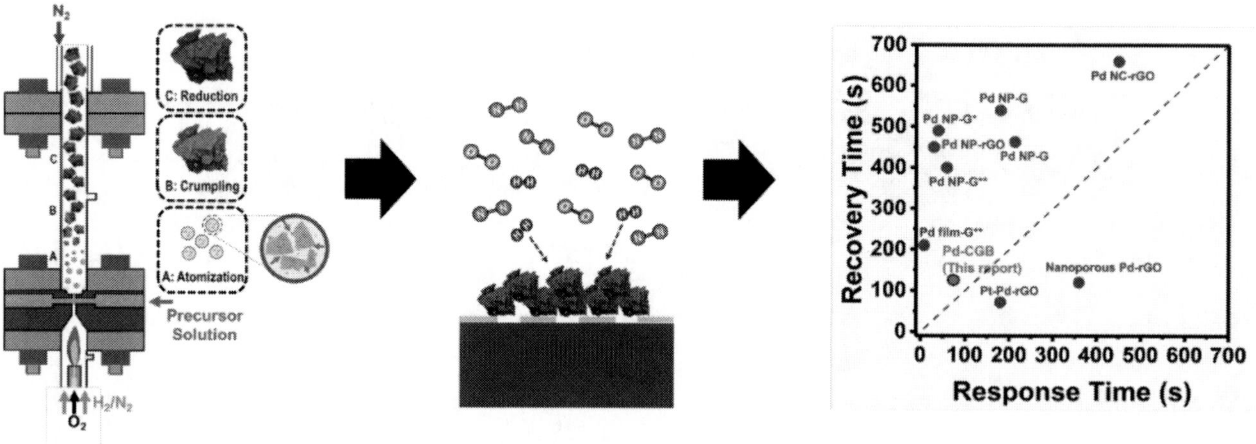

601668 Carbon Black-Gold Nanoparticles for Detection of Analytes Using Surface Enhanced Raman Scattering

Abstract:

Surface-enhanced Raman scattering (SERS) is a powerful technique for the detection of a variety of molecules at very low concentrations. We have developed irregularly shaped gold nanoparticles using a carboxyl-terminated carbon black (CB) nanoparticles as a template. The CB particles were coated with the cationic polyelectrolyte poly-L-lysine (PLL) and exposed to a tetrachloroauric acid solution. Gold - carbon black (AuCB) particles were formed by the reduction of gold chloride ions that concentrated on the surfaces of the PLL. The morphology of the gold on the AuCB can be modified using different PLL concentrations,. The plasmon resonance can then be matched to the incident laser wavelength, thus greatly enhancing Raman signals coming from analytes in the vicinity of the gold surface. We used these AuCB particles to detect a wide range of analytes, including 4-nitrobenzene thiol, and nitrate and sulfate ions in a mixture. The limits of detection were well below that obtained by other SERS particles.

607107 Local Curvature Enables Spatially-Defined Properties in Graphene

Abstract:

Selectively functionalized graphene can realize spatially-defined properties that are highly desirable for atom-thin devices. Because curvature tunes the reactivity of graphene, patterning graphene into regions of different local curvatures can achieve domains with different levels of functionalization. Previous buckling methods were limited by the range of tunability over curvature and control of the orientation at the microscale. This presentation describes a scalable approach to achieve spatially selective graphene fluorination using multiscale wrinkles. Graphene wrinkles were formed by relieving the strain in thermoplastic polystyrene substrates conformally coated with fluoropolymer and graphene skin layers. Chemical reactivity of a fluorination process could be tuned by changing the local curvature of the graphene nanostructures. Patterned areas of graphene nanowrinkles and crumples followed by a single-process plasma reaction resulted in graphene having different fluorination levels side-by-side. Because the reduction in conductivity scales with the degree of fluorination, the conductivity of the graphene can now be controlled as a function of feature size.

611391 Nir-Fluorescent Carbon Nanotube Sensors: Form Factors and Applications

Abstract:

Colloidal single-walled carbon nanotube (SWCNT) optical sensors are capable of highly sensitive and selective molecular recognition for varying classes of analytes. However, their incorporation into commercial uses requires the study of their packaging or form factors necessary for specific applications. Here, we have showcase two techniques that are steps toward incorporating these materials for both industrial and field applications. First, we study the immobilization of SWCNT sensors on seven different types of paper substrates. Sensors remain functional upon immobilization for up to 60 days with storage technique dictating sensor longevity. With a wax printing method, we are able to form one-dimensional barcodes that allow for rapid reading of sensor elements. Using a sensor array of ssDNA wrappings, we are able to differentiate in test solutions between Cu(II), Cd(II), Hg(II), Pb(II) at a concentration of 100 μM.

Next, we design and characterize a compact, integrated fiber optic nanosensor probe element (optode) for pharmaceutical applications. The smaller form factor of the optode system is designed to facilitate incorporation into on-line or at-line biopharmaceutical processes. The sensor tip itself consist of fibers that provide both excitation and emission. Together with a 3D printed capsule, we embed SWCNT sensors in a protective and porous hydrogel for the sampling of a production stream. By adjusting the crosslinking density of the hydrogel, we are able to extract time dependent sensor responses as a function of analyte size. Using this platform with a SWCNT sensor sensitive to IgG, we are able to resolve the relative fraction of monomers versus dimer aggregates in under 5 minutes of analysis.

610030 1D Fiber Spinning, 2D Coating Technology and 3D Printing for Nanocarbon-Included Composite Manufacturing

Abstract:

This research will describe a few manufacturing methods incorporating nanoscale carbons in polymeric nanocompsoites. These composites are useful for either structural composites or funcitoanl hybrid materials. The polymers and carbon nanometerials will form layered strucutres in 1D fibers, 2D coatings and 3D stacked laminates. Layer-structured composites have a broad application in light composites, capacitors, drug delivery, batteries, sensors, actuators, and many others. Here reported are three case studies of layered structures at nano- and micro-scales that display great potential in rapid prototyping of versatile sensors. The first example exhibited the use of a few polymers and nano carbons in co-axial composite fibers. This demonstraiton leveraged the tooling engineering of the spinneret and successfully fabricated multiphase composite fibers that can simultaneously exfoliate and align graphene layers. The second exhibition was about the layeres of conjugating polymers and carbon nanoparticles, especially about how the nanostructure will influence the thermal and thermoelectrical perforances. The last case was the feasibility of combining 3D printing and dip coating in making layered composites. During the nanoparticle assembly procedure, the surface morphology and the solid-liquid-air interface control exhibited efficient management of nanoparticle topologies for a unit layer.

611338 Entirely Water-Based Ultralight Graphene Fibers with Superior Properties

Abstract:

Fabrication of graphene fibers with high thermal and electrical conductivity, and good mechanical properties via facile processing method is yet a challenge for industries. The common fabrication of graphene oxide fibers originated from graphite, requires several steps of oxidation and reduction and still, the exceptional thermal and electrical conductivity and stiffness of the graphene sheet as a single layer, has not been translated in such fibers.

Towards this goal, we have fabricated ultralight graphene fibers with the density of 0.25 g/cm^3, and an exceptional thermal and electrical conductivity of 2Í10^3 W/K-m and 2Í10^5 S/m, respectively. With a ~0.5 GPa strength and 40 GPa Young's modulus, the GFs possess the three key element properties of high thermal, electrical conductivity and stiffness. Furthermore, being entirely water-base and by addition of a mild heat-treatment at 600 °C, the fabricated GFs in this work are promising candidates for industrial light-weight composites applications. The inclusion of such fibers in epoxy composites lead to floating properties that is of interest of versatile industries.

604859 DNA-Wrapped Boron Nitride Nanotubes in Alcohol/Water Mixture Systems: Experimental and Simulation Study

Abstract:

Aligned films of boron nitride nanotubes (BNNTs) have potential applications in thermal interface materials and electronic and optoelectronic devices due to their promising optical, mechanical, and thermal properties. We previously reported dispersions of BNNTs *via* probe tip sonication in aqueous solutions of sequence-controlled single-stranded DNA. However, the harsh ultrasonication may cause defects and cutting of nanotubes, which fundamentally affect final properties of their macroscopic assemblies. In this work, we demonstrated a simple and scaled-up dispersion process to produce DNA-wrapped BNNT hybrids (DNA-BNNTs) using a cheaper, natural DNA source by a mild bath sonication in various alcohol/water mixture systems including methanol, ethanol, and isopropyl alcohol. Molecular dynamics simulations were utilized to better understand the solvation behaviour of BNNTs in a range of alcohol/water mixtures. Moreover, length distributions of BNNTs dispersed in alcohol/water mixtures were characterized by rheology and SEM imaging to better estimate the apparent length distributions of nanotubes. Our results elucidated the scalable and efficient dispersions of BNNTs which plays a vital role in fundamentally improving the final properties of macroscopic assemblies of BNNTs.

608142 Development of Supervised Learning Models for Protein Adsorption to Engineered Nanoparticles

Abstract:

Engineered nanoparticles are an advantageous platform for developing sensors and delivery vehicles for numerous biotechnology applications[1, 2]. Most such technologies are developed and validated *in vitro* but intended for use in complex biological environments *in vivo*. To-date, testing the compatibility of nanotechnologies in biological systems requires a heuristic approach, whereby unpredictable biofouling and off-target effects often prevent implementation of nanotechnologies. Such biofouling effects are frequently the result of spontaneous adsorption of proteins to the nanoparticle surface, forming the 'protein corona' and altering the physiochemical properties, and thus intended function, of the nanotechnology. To implement engineered nanoparticles in biological applications, we require predictive power over which proteins and to what extent they preferentially adsorb to the nanoparticle surface, in different biofluids, to form the protein corona.

In this work, we develop a random forest classifier (RFC) trained using mass-spectrometry data to model which proteins populate the corona of a candidate optical nanosensor composed of single-walled carbon nanotubes, SWNTs, functionalized using single stranded DNA molecules, $(GT)_{15}$. $(GT)_{15}$–SWNTs were incubated in biofluids relevant to their *in vivo* applications – blood plasma and cerebrospinal fluid – to populate the nanoparticle's protein corona. Proteins adsorbed in the corona phase were purified then identified and quantified via proteomic mass spectrometry. Using publicly available protein databases, the RFC was trained to predict the propensity for specific proteins to populate the corona phase from each biofluid. We expand upon this work by investigating a panel of other relevant nanoparticles, including multiple functionalizations of SWNTs, polystyrene nanoparticles, and gold nanoparticles, evaluating their protein corona compositions in blood plasma, cerebrospinal fluid, and plant lysate. Moreover, we include additional classification techniques to understand protein physicochemical and biological factors important in predicting selectively adsorbed proteins. Lastly, we apply the results of this model to rapidly identify candidate proteins with high binding affinity to different nanoparticles, and experimentally validate adsorption of these proteins to DNA-SWNTs with a fluorescence-based corona exchange assay to evaluate the predictive power of the model[3]. The classifier presented here provides an important tool for predicting protein-nanoparticle interactions, which is needed for effective translation of bionanotechnologies from *in vitro* synthesis to *in vivo* use.

References

(1) Beyene, A. G.; Delevich, K.; Del Bonis-O'Donnell, J. T.; Piekarski, D. J.; Lin, W. C.; Wren Thomas, A.; Yang, S. J.; Kosillo, P.; Yang, D.; Prounis, G. S.; Wilbrecht, L.; Landry, M. P. Imaging Striatal Dopamine Release Using a Nongenetically Encoded near Infrared Fluorescent Catecholamine Nanosensor. *Sci. Adv.* **2019**, *5* (7), eaaw3108.

(2) Demirer, G. S.; Zhang, H.; Matos, J. L.; Goh, N. S.; Cunningham, F. J.; Sung, Y.; Chang, R.; Aditham, A. J.; Chio, L.; Cho, M. J.; Staskawicz, B.; Landry, M. P. High Aspect Ratio Nanomaterials Enable Delivery of Functional Genetic Material without DNA Integration in Mature Plants. *Nat. Nanotechnol.* **2019**, *14* (5), 456–464.

(3) Pinals, R. L.; Yang, D.; Lui, A.; Cao, W.; Landry, M. P. Corona Exchange Dynamics on Carbon Nanotubes by Multiplexed Fluorescence Monitoring. *J. Am. Chem. Soc.* **2020**, *142* (3), 1254–1264.

608272 A Database of Porous Rigid Amorphous Materials

Abstract:

The development of publicly available materials databases for materials including zeolites (IZA-SC), metal organic frameworks (CoRE-MOF), and inorganic materials (The Materials Project) has been a key enabler of high-throughput computational screening and data-driven discovery of materials for potential use in new technologies. This work aims to develop a similar database for structures of porous rigid amorphous materials, an important class of materials for which no such resource is currently available. The database includes atomically detailed structures of disordered materials like amorphous carbons, kerogens, hyper-cross-linked polymers etc. generated using a wide range of simulation techniques by multiple research groups. We present extensive computational analyses for material characterization by calculating a series of scalar (e.g. accessible surface area) and vector (e.g. pore size distribution) descriptors. In addition to structural descriptors, a variety of gas adsorption isotherms for both single component and binary mixtures are predicted for each structure. We also discuss the agreement between binary adsorption data and predictions from the Ideal Adsorbed Solution Theory.

608545 Optimal Design of Novel Precursor Material Using Camd for Enhanced Growth Kinetics of ALD

Abstract:

Optimal design of novel precursor material using CAMD for enhanced growth kinetics of ALD

Rajib Mukherjee[1,2], Mina Shahmohammadi[3], Christos G. Takoudis[3,4], Urmila M. Diwekar[2, 4*]

1. *Department of Chemical Engineering, University of Texas Permian Basin, Odessa, TX 79762*
2. *Vishwamitra Research Institute, Crystal Lake, Illinois, 60012*
3. *Department of Chemical Engineering, University of Illinois at Chicago, Chicago, Illinois 60607*
4. *Department of Bioengineering, University of Illinois at Chicago, Chicago, Illinois 60607*

**Corresponding: urmila@vri-custom.org*

Abstract

Atomic Layer Deposition (ALD) is a promising technique in which deposition of thin films takes place on a substrate through sequential and self-limiting surface reactions. A precursor used in this process chemisorbs on the substrate and part of the molecule subsequently desorbs from the surface after completion of the reaction leaving behind deposition of metal and metal oxides on the substrate. Precursor chemisorption on the substrate leads to a self-limiting process and it eventually results in films with desired thickness at Ångström length scale. In order to design an ALD experiment, one should decide on precursor(s) based on the ALD conditions (i.e., bubbler and reactor temperatures, pressure, gas flow, etc.) and likely applications of the final film. It is practically impossible to carry out a huge number of ALD experiments using numerous precursors and deposition conditions in order to find the optimum one depending on the applications of interest. Moreover, only existing precursors can be tested for experiments. In this work, we will present development of novel titanium precursors from the thermodynamic model of adsorption as in Adsorbate solid solution theory (ASST) using Computer Aided Molecular Design (CAMD) [1]. Three different functional groups are used for the experiment, the thermodynamic properties of which are obtained through group contribution method (GCM) [2]. Each functional group is allowed to vary from 0 to 20 times. Since this results in combinatorial explosion of alternatives, a combinatorial optimization method is applied using novel efficient ant colony optimization (EACO) algorithm [3]. This is the first time CAMD is being applied to design precursor materials for ALD. The ALD growth kinetics is used as an objective of optimization and a solution is validated with thermodynamic constraints. From our simulation, 41 different novel precursor molecular structures are generated with growth rates ranging from 1.23 Å/cycle to as high as 1.65 Å/cycle. Thus, several of these novel precursors have shown growth rates higher than the known titanium precursors [4]. Temperature as a variable with wide range of limits between 300 K and 600 K is used for our simulation. ALD growth rate is found to be a function of the combination of the precursor functional groups as well as temperature with a complex correlation among them.

Keywords: Atomic Layer Deposition, Computer-Aided Molecular Design, Group Contribution Method, Adsorbate Solid Solution Theory, ALD growth kinetics

[1] Berti, C., Ulbig, P., Burdorf, A., Seippel, J., Schulz, S., 1999. Correlation and prediction of liquid-phase adsorption on zeolites using group contributions based on adsorbate-solid solution theory. Langmuir 15, 6035–6042. https://doi.org/10.1021/la981415p

[2] M. Shahmohammadi, R. Mukherjee, C.G., Takoudis, U.M., Diwekar, 2019, "Optimal Design of Novel Precursors Materials for the Atomic Layer Deposition. Part 1: Group Contribution Method", In preparation

[3] Gebreslassie, B. H., & Diwekar, U. M. (2015). Efficient ant colony optimization for computer aided molecular design: Case study solvent selection problem. Computers & Chemical Engineering, 78, 1-9.

[4] R. Mukherjee, M. Shahmohammadi, C.G., Takoudis, U.M., Diwekar, 2019, "Optimal Design of Novel Precursor Materials for Atomic Layer Deposition. Part 2: Computer-Aided Molecular Design", In preparation

608582 Nature-Inspired Self-Cleaning Surfaces: Mechanisms, Modelling, and Manufacturing

Abstract:

Nature-inspired self-cleaning surfaces have attracted considerable attention from both fundamental research and practical applications. Around 2005, the number of publications on this topic started strongly rising, due to the increased availability of nanoscale manufacturing techniques. In this contribution, we will adopt a chemical-engineering point of view and give an overview of mechanisms, modelling, and manufacturing (M3) of state-of-the-art, nature-inspired self-cleaning surfaces. We will introduce six nature-inspired self-cleaning mechanisms: The Lotus-effect, superhydrophobic-induced droplet jumping, superhydrophobic-induced unidirectional movement of water droplet, underwater-superoleophobic-based self-cleaning, slippery-based self-cleaning, and dry self-cleaning. These mechanisms of nature self-cleaning examples are popular and well-known as well as have been widely applied or exhibited potential applications in our daily life and industrial productions. The mathematical and numerical modelling of the identified self-cleaning mechanisms will be carefully introduced, which will contribute to the rational design and reproducible construction of these functional self-cleaning surfaces [1].

We will also discuss how these materials can be produced, with a focus on scalable manufacturing. There are already some tens of companies offering commercial self-cleaning coatings, and their number grows every year. The major bottleneck to bring self-cleaning surfaces on the market remains the lack of durability. Coatings usually do not retain non-wetting properties, due to mechanical abrasion, UV degradation or organic contamination. To convince end-user companies to incorporate self-cleaning coatings into their products (e.g. shoes, cars, construction materials) remains a challenge, and likely requires a scientific breakthrough in mechanical durability. Further progress in this field will require steps in understanding, which would profit from interdisciplinary collaboration.

[1] Yu, C., Sasic, S., Liu, K., Salameh, S., Ras, R. H. A., & van Ommen, J. R. (2020), Chem. Eng. Res. Des. 155 (2020), 48-65.

611823 Accelerated Prediction of Atomically Precise Cluster Structures Using on-the-Fly Machine Learning

Abstract:

The chemical and structural properties of atomically precise nanoclusters are of great interest in numerous applications, but predicting the structures of the clusters can be computationally expensive. In this work, we present a procedure that can rapidly predict low-energy nanocluster structures by combining a genetic algorithm and an interatomic potential model actively learned on-the-fly. A pool-based genetic algorithm is implemented to efficiently sample the configuration space, and moment tensor potentials are used to rapidly relax and evaluate the energies of predicted clusters. The resulting procedure significantly accelerates the process of identifying low-energy cluster structures and is demonstrated on both bare and ligated clusters. The predicted lowest-energy nanoclusters are compared with the lowest energy structures reported in literature to validate this methodology. This workflow provides a feasible way to systematically predict low-energy structures for nanoclusters at a large scale, which can greatly facilitate the discovery and design of novel nanomaterials for a wide range of applications.

601876 Metal-Free Carbon Electrocatalysts for Efficient Energy Conversion and Storage

Abstract:

Green and renewable energy technologies, such as fuel cells, batteries, and water-splitting systems, hold great promise to solve current energy and environmental challenges. However, noble metal catalysts (*e.g.*, Pt, Pd, RuO_2, IrO_2) are generally needed to promote the *hydrogen evolution reaction* (HER) for hydrogen fuel generation from photo-electrochemical water-splitting, *oxygen reduction reaction* (ORR) in fuel cells for energy conversion, and *oxygen evolution reaction* (OER) in metal-air batteries for energy storage. The high cost of precious metal-based catalysts and their limited reserve have precluded these renewable energy technologies from large-scale applications.

Along with the recent intensive research efforts in non-noble metal based ORR catalysts, we have previously demonstrated that vertically-aligned nitrogen-doped carbon nanotubes could actively catalyze ORR *via* a four-electron process free from the CO poisoning effects with a 3-time higher electrocatalytic activity and better long-term durability than that of commercial Pt/C catalysts. The improved catalytic performance was attributed to the doping-induced charge transfer from carbon atoms adjacent to the nitrogen atoms to change the chemisorption mode of O_2 and to readily attract electrons from the anode for facilitating the ORR.

Subsequently, it was demonstrated that various graphitic carbon materials, doped with heteroatoms of different electronegativities from that of carbon atom, physically adsorbed with certain polyelectrolytes, and even without any apparent dopant or physically adsorbed polyelectrolyte, could also exhibit good ORR performance. More recent studies have further demonstrated that certain heteroatom-doped carbon nanomaterials could act as metal-free bifunctional catalysts for ORR/OER in metal-air batteries for energy storage, and even ORR/OER/HER trifunctional catalysts for self-powered water-splitting to generate hydrogen fuel and oxygen gas from water.

In this talk, I will summarize some of our work on the metal-free catalysts based on carbon nanomaterials for various energy-related reactions, along with an overview on the recent advances and perspectives in this exciting field.

602186 Electrospun Particle/Polymer Fiber Mats As Fuel Cell and Battery Electrodes

Abstract:

Electrospinning is gaining popularity as a convenient, robust, and scalable technique for fabricating non-woven mats of sub-micron diameter polymer fibers. Although not as well studied, the technique can also be used to prepare particle/polymer fiber networks with high intra- and inter-fiber porosity. Such fibrous mats can be used as porous electrodes in fuel cells and batteries, where a high electrode/electrolyte interfacial area is needed. In the present talk, recent experimental work on new nanofiber mat compositions and architectures for hydrogen/air fuel cell cathodes and lithium-ion battery electrodes will be presented. The fuel cell work is focused on high particle content fibers containing Pt/C or Pt-alloy catalyst powder and a polymer binder based on Nafion® perfluorosulfonic acid. The fuel cell power output using such a cathode is very high for a low Pt loading (0.10 mg/cm^2), near 1.0 W/cm^2 for H_2/air feeds. Additionally, the fiber mat cathodes exhibit excellent durability, as determined by power losses after accelerated voltage cycling tests. For Li-ion battery applications, single and dual fiber anode mats have been prepared and evaluated, where the fibers contain a polymer binder (e.g., poly(acrylic acid), polyamidimide, or polyvinylidene fluoride), Si nanoparticles as the electrochemically active material, and carbon powder as an electron conductor. Here, the gravimetric and areal capacities of the fiber anode are considerably higher than those for a conventional slot-die coated slurry anode. Inter-fiber voids allow for volume changes in Si during lithiation/delithiation, thus minimizing capacity fade during charge/discharge cycling (up to 1,000 cycles). Experimental details for both fuel cell and battery electrodes will be presented in this talk, including electrospinning procedures, fiber mat characterization data, and device performance.

611471 Bottom-up All Aqueous Assembly Approach for the Synthesis of Lightweight 3D Carbon Nanocomposite Aerogels for Electrochemical Energy Storage and Conversion Applications

Abstract:

Novel Bottom-Up All Aqueous Assembly Approach for the Synthesis of Lightweight 3D Carbon Nanocomposite Aerogels for Electrochemical Energy Storage and Conversion Applications

Brigit A. Duffy[1], Gabrielle M. Milanesa[1], An B. Vu,[1] Jordan M. Davis[1], Duncan R. Day[1], Pamela L. Sheehan[2], Preston C. Haney[2], Harry L. Moore Jr[2], F. John Burpo[1], Enoch. A. Nagelli[1*]

[1]Department of Chemistry & Life Science, Chemical Engineering Program, United States Military Academy, West Point, New York 10996

[2]U.S. Army Combat Capabilities Development Command, Army Futures Command, CCDC Armaments Center, Picatinny Arsenal, New Jersey 07806

Carbon nanomaterials constitute an ideal miniaturized composite platform for integration with other solid-state electrode materials. Self-assembly has been recognized as an effective strategy for the bottom–up synthesis of 3D macrostructures using graphene and CNTs as building blocks. [1, 2] We demonstrate a novel, all-aqueous, and scalable 3D platform materials design process for the synthesis of carbon nanocomposite aerogels to serve as porous ultralightweight electrodes for the next generation energy storage and conversion applications. Specifically, the negatively charged surface functional groups on graphene oxide and oxidized CNTs will be the sites for electrostatic coordination of positively charged noble metal cations and polyelectrolytes from aqueous solutions. This enables the incorporation of carbon nanomaterials with any polyelectrolyte and noble metal nanostructures with precise connection of each of the individual components while maintaining their unique properties.[3-5] Through solution-based self-assembly followed by critical point drying and chemical reduction, we can develop hierarchical porous 3D nanocomposite aerogels comprised of carbon nanomaterials with any precious or noble metal nanostructures (Pt, Pd, Au, Ru, Ag, and Ru) and polyelectrolyte. Electrochemical techniques such as cyclic voltammetry, linear sweep voltammetry, galvanostatic charge-discharge, and impedance spectroscopy are used to determine the electrical and ionic conductivity, electrocatalytic activity, and supercapacitor performance of the aerogels as lightweight high-power and energy density electrodes.

KEYWORDS: Three-dimensional Nanomaterials, Nanocomposites, Aerogels, Electrocatalysts, Batteries, Fuel Cells

CONTACT: Enoch A. Nagelli, Email: enoch.nagelli@westpoint.edu

References

[1] L. Dai, “Intelligent Macromolecules for Smart Devices: From Materials Synthesis to Device Applications”, Springer: Berlin, 2004.

[2] L. Dai, (Ed.) "Carbon Nanotechnology: Recent Developments in Chemistry, Physics, Materials Science and Device Applications", Elsevier: Amsterdam, 2006.

[3] M. Yanga, Y. Houd, N. A. Kotov, "Graphene-based multilayers: Critical evaluation of materials assembly techniques" Nano Today 2012, 7, 430.

[4] E. Nagelli, R. Naik, Y. Xue, Y. Gao, M. Zhang, and L. Dai "Sensor arrays from multicomponent micropatterned nanoparticles and graphene" Nanotechnology 2013, 24, 444010.

[5] E. Nagelli, L. Huang, A. Q.-Z. Dai, F. Du, L. Dai, "3D Vertically-Aligned CNT/Graphene Hybrids from Layer-by-Layer Transfer for Supercapacitors" Part. Part. Syst. Charct. 2017, 34, 1700131. Article in Special Issue of Graphene Oxide Liquid Crystals.

[6] F. J. Burpo, E. A. Nagelli, L. A. Morris, J. P. McClure, M. Y. Ryu, J. L. Palmer. "Direct Solution-Based Reduction Synthesis of Au, Pd, and Pt Aerogels." J. Materials Research. 2017, 32 (22), 4153-4165. (Invited Paper) Featured in MRS Bulletin: "Metallic aerogels synthesized in one-step process," December 18, 2017.

610947 Tailoring Internal Pores for Mitigation of Volume Expansion in Silicon Graphene Hybrid Anodes for Lithium Ion Batteries

Abstract:

The current state of art Lithium ion batteries (LIB) are unable to satisfy the demands for electric vehicle (EV) applications due to their low energy density. To improve their energy density, silicon as anode material has been very promising due its extraordinary theoretical capacity. However, silicon undergoes huge volume expansion(~ 400%) during lithium-ion storage. This leads to the pulverization of the electrode structure over several cycles of lithiation and delithiation and consequently severe capacity decay. Thus, mitigation of this volume expansion has been a stumbling block for its commercialization. One approach to address this issue is to use silicon carbon-based composites and engineering void spaces for the effective accommodation of silicon volume expansion. The internal pores give breathing room for the silicon to expand, which maintains the structural integrity. Many scientists have researcher methods to mitigate the silicon volume expansion by engineering voids; however, there is a lack of direct correlation in terms of the size of pores created and the mitigation of volume expansion achieved. Additionally, the various approaches studied in the literature to create pore spaces invariably involve the use of expensive fabrication methods.

In our study we are using a two-step approach to form tailored internally porous silicon graphene hybrid anodes. In the first step, we form directly deposited silicon graphene hybrid anodes via air-controlled electrospray, and in the second step we use a mild heat treatment to thermally remove poly acrylic acid, which serves as our sacrificial template. This approach provides structural integrity and volume expansion accommodation. In order to prove our hypothesis of mitigation of silicon volume expansion via the creation of internal pores, we also study in-situ volume expansion via a dilatometer and compare the resulting mitigation of volume expansion for Si/Gr with pores and Si/Gr without pores. Our conclusions show the interplay of creation of pores and the mitigation of volume expansion over a number of charge/discharge cycles.

610999 Scalable Surface Application of Stabilized Lithium Metal Particles (SLMP) for Enhancing Silicon-Based Anode in Full-Cell Lithium-Ion Batteries

Abstract:

Silicon has been considered as one of the most promising replacements for graphite anodes in next-generation lithium-ion batteries due to its superior energy density. However, one prominent issue with silicon-based anode materials is the substantial consumption of lithium ions due to the large volume expansion of silicon upon lithiation and the continuous formation of the solid-electrolyte interphase. In a full-cell setting where silicon is paired with a commercial cathode, this irreversible consumption of lithium ions is especially detrimental to the limited lithium-ion reserve of the system, leading to a low first-cycle Coulombic efficiency and a shortened cycle life. Several pre-lithiation strategies have been explored to mitigate this problem, and the use of stabilized lithium metal particles (SLMP) is being demonstrated as a practical pre-lithiation method compatible with scaled-up industrial applications. Particularly, a SLMP slurry was designed with hexanes as a compatible solvent and a nonionic surfactant additive at 1%wt serving as the stabilizer for maintaining a uniform SLMP suspension. This slurry was then spray-coated onto the prefabricated silicon electrode, forming a uniformly distributed and well adhered SLMP layer on the surface. By optimizing the amount of SLMP being applied, we are able to demonstrate improvements in the electrochemical performance of silicon-based full-cell batteries with both increased first-cycle Coulombic efficiency and capacity retention. We will also attempt to provide a description of SLMP's effects at the electrode surface throughout the electrochemical cycling of the batteries.

610144 Corn Stover-Derived Biocarbon from Hydrothermal Liquefaction for Supercapacitor Applications

Abstract:

Biocarbon for supercapacitor anodes has become a frontrunner for renewable energy storage materials. Its unique physical characteristics allow for enhanced rate capabilities, increased ion transport, all leading to sustainable scale-up. Current work focuses on the use carbon-rich residue from the corn stover refining. Value added chemicals were first extracted by enzymatic hydrolysis, and the resultant lignin-rich unhydrolyzed solids (UHS) derived from preprocessing of corn stover were subjected to hydrothermal liquefaction (HTL). Bio-char samples derived from investigating the HTL reaction kinetics at various temperatures, initial purge, reaction time, and biomass loading, were subjected to thermal treatment at 400 °C under nitrogen for 2 h and followed by the activation at 850 °C under Ar for 3 h to obtain the biocarbon material. Electrochemical results show that select samples reach capacitances above 200 F g^{-1} at 5 mV s^{-1}. Physical and electrochemical characterizations such as BET, XRD, SEM, and ATR-FTIR, Raman, cyclic voltammetry, and chronopotentiometry were carried out to fundamentally probe the biocarbon. This work provides an insight into how pretreatment conditions affects the performance of biocarbon for supercapacitor applications.

608709 High Biological Drug-Loaded Calcium Phosphate Nanocarriers Produced By Flame Spray Pyrolysis

Abstract:

Nanoparticle-based drug delivery systems have been considered as an auspicious alternative for efficient delivery of biological drugs (such as proteins and peptides) [1], which is deemed challenging due to large size and enzymatic degradation susceptibility of biologics. However, the clinical translation of such therapeutic nanoformulations is still in its infancy due to (i) complex synthesis methods of nanocarriers, and (ii) poor loading efficiency of biologics in comparison to clinically relevant therapeutic doses [2].

Here, we aim to address these challenges by engineering calcium phosphate (CaP) nanocarriers with tailored properties by flame spray pyrolysis (FSP) [3,4]. FSP is a scalable process with high production rates and also allows for fine tuning of nanoparticles properties. CaP nanoparticles with different specific surface areas (SSA) are obtained by varying precursor concentration in the flame. Our main goal is the synthesis of nanoparticles with high loading capacity of biological drugs (proteins and peptides). Bovine serum albumin (BSA) and bradykinin are used as model protein and peptide, respectively, and are physisorbed on the as-prepared nanoparticles. Their concentration along with the incubation time are varied in order to investigate their influence on loading efficiency. High loading values, reaching 350 mg/g and 600 mg/g of CaP for BSA and bradykinin, respectively, are obtained.

Furthermore, as a proof of concept, we apply the developed protocol and physisorb LL-37 antimicrobial peptide on CaP nanoparticles. Maximum loading achieved here (~800 mg/g) outperforms all LL-37 loading values found in the literature both for inorganic and organic nanocarriers [5]. Typical fractal-like structure of flame made agglomerated/aggregated nanoparticles promotes high drug loading efficiency. Moreover, it is shown that CaP nanoparticles protect LL-37 from enzymatic degradation by Protease K *in vitro* and that the peptide retains its antimicrobial activity against *Escherichia coli* and *Streptococcus pneumoniae* after being physisorbed on nanoparticles. The antimicrobial activity of LL-37 loaded nanoparticles, as assessed by monitoring bacterial growth, outperforms that of the free peptide for *E. coli*, whereas similar behavior is observed for free LL-37 and LL-37 loaded CaP nanoparticles for *S. pneumoniae*.

In a nutshell, the potential of FSP for fabrication of nanocarriers with high drug loading capacity of biologicals is shown in this study, which will improve therapeutic efficiency by delivering high concentration of biological drugs at the diseased sites.

References

1. Soares S., Sousa J., Pais A., Vitorino C., Nanomedicine: Principles, Properties, and Regulatory Issues. *Chem.* 2018;6:360.
2. Hua S., de Matos M.B.C., Metselaar J.M., Storm G., Current Trends and Challenges in the Clinical Translation of Nanoparticulate Nanomedicines: Pathways for Translational Development and Commercialization. *Pharmacol.* 2018;9:790.
3. Mädler L., Kammler H.K., Mueller R., Pratsinis S.E., Controlled synthesis of nanostructured particles by flame spray pyrolysis. *Aerosol Sci.* 2002;33:369-389.
4. Loher S., Stark W.J., Maciejewski M., Baiker A., Pratsinis S.E., Reichardt D., Maspero F., Krumeich F., Günther D., Fluoro-apatite and Calcium Phosphate Nanoparticles by Flame Synthesis. *Mater*. 2005;17:36-42.
5. Tsikourkitoudi V., Karlsson J., Merkl P., Loh E., Henriques-Normark B., Sotiriou G.A., Flame-made calcium phosphate nanoparticles with high drug loading for delivery of biologics. *Molecules* 2020;25:

597503 Efficient Intracellular Delivery of CRISPR Payloads Mediated By a Polymeric Vehicle Discovered through Combinatorial Design and High-Throughput Experimentation

Abstract:

The evolution of genome editing strategies such as CRISPR/Cas9 is rapidly transforming genetic research and personalized medicine. Despite the emergence of sophisticated gene editing tools with increasing specificity and safety, clinical translation has been hindered by challenges in engineering delivery vehicles. Vehicles must juggle multiple functions: package multiple payloads into nanostructures that undergo efficient cellular internalization, provide storage stability and maintain payload integrity by preventing degradation, and finally release the payload at the target tissue within a desirable spatiotemporal window. While current viral vectors, including adenoassociated viruses (AAVs), have provided clinical success for treating several diseases, the astronomical costs per dose ($0.4M–$2.1M for current FDA-approved therapies) is likely to cause immense financial strain on the healthcare system. These costs are driven in part by the expensive and lengthy processes to optimize and mass-manufacture clinical grade viral carriers. Moreover, the limited cargo capacity of viral carriers and immune responses provoked at higher doses has driven this field towards synthetic substitutes that offer lower costs, scalability, and safer clinical outcomes for repeat administration.

The current paradigm of polymer design involves inefficient empirical approaches centered on synthesizing, characterizing and evaluating "one-structure-at-a-time" . On the other hand, *ab initio* prediction of gene editing efficiency directly from polymerl structure faces roadblocks due to limited mechanistic understanding. To accelerate *de novo* discovery of polymeric vectors, we must diversify the pool of candidate materials and build statistical frameworks that systematically capture biophysical trends correlating structure, properties, and performance in disease-relevant tissues. To this end, a chemically diverse library of well-defined statistical copolymers was synthesized and complexed with clinically relevant genetic payloads such as plasmid DNA and ribonucleoprotein complexes (RNP). These formulations were subjected to high-throughput screening assays to identify hit polymers for each payload. The hit polymer was benchmarked against commercial reagents marketed for RNP delivery, an was found to outperform state-of-the-art reagents such as Lipofectamine 2000, Lipofectamine CRISPRMAX, JetCRISPR and JetPEI. While Lipofectamine CRISPRMAX achieved editing in only 26% of cells, our hit formulations resulted in gene editing in up to 58% of cells. Further, preliminary data indicates that in addition to facilitating gene knockouts via non-homologous end joining (NHEJ), our polymer can co-deliver donor DNA plasmids with the RNP payload to achieve gene knock in via homology-directed repair (HDR) pathways.

Subsequently, we deployed statistical tools to identify the structural drivers for cellular internalization of RNP payloads, cell viability and transfection efficiency and mined structure-function correlations. These predictive relationships shed light on the critical role played by hydrophobic design motifs in mediating efficient gene editing and can aid experimentalists engaged in polymer synthesis for diverse gene therapy applications .This data-driven approach has yielded an efficient vector discovery pipeline and led to the identification of a delivery system that does not conform to traditional heuristics.

602407 Delivery of CRISPR/Cas9 Gene-Editing Systems By Cell-Penetrating Magnetite Vehicles: Synthesis, Characterization and *in Vitro* Testing

Abstract:

Genetic diseases currently impact about 7% of the world's population [1]. One of the alternatives to treating these diseases is gene therapy. Such therapies are becoming increasingly important in the field of bioengineering as they allow modification of faulty or altered genetic material by either replacing, reducing or suppressing the expression of genes involved in the disease [2]–[4].

One of the main challenges in gene therapy is the transport of genetic material into the cells of interest since a number of obstacles need to be overcome including rapid degradation of genetic material by the physiological environment, limited cell uptake and poor endosomal escape. As a result, few alternatives have been implemented to tackle these issues such as the use of viral vectors, and the use of nanostructured materials [5]–[7]. Due to biosafety concerns, using viral vectors has been questioned, while some of the nanostructured materials have shown lower delivery efficiencies [8]–[10].

To improve the efficacy of the delivery nanomaterials capable of responding to magnetic fields are used. This approach is known as magnetofection and relies on using magnetic fields to guide the delivery of the nanovehicles [11]–[14]. Using magnetic nanomaterials has attracted significant attention for applications in biomedicine due to their ease of handling, their ability to transport various types of biomolecules, and the possibility of controlling their fate to specific tissues and organs [15]–[17]. Consequently, the chances for side effects can be significantly reduced [18], [19].

One of the most appealing magnetic nanomaterials is magnetite, which has gained attention over the past decades thanks to its applicability as cell marker, active biologicals separation, drug delivery, and as contrast agent in MRI testing [20]–[23]. This could be attributed to their high biocompatibility and proven possibility of renal excretion as ferritin [21], [24], [25].

To enable the delivery of gene therapies via nanomaterials, the genetic material needs to be covalently conjugated to the surface of the nanomaterials with using highly-reactive functional groups such as thiol, amine, hydroxyl or carboxyl [26]. Additionally, nucleotide cargoes should be able to be released intracellularly and ideally reaching the nucleus. For this purpose, a number of chemistries have been incorporated into the delivery vehicles that respond to environmental changes or stimuli making the system smart [25]. With the emergence of CRISPR/Cas9 systems, more robust and complex interventions can be achieved, and consequently larger nucleotide sequences need to be delivered efficiently, in a cost-effective manner [9], [27].

As abovementioned, the objective of this study is to develop a nanostructured platform for immobilization, and intracellular release of nucleic acids with application in gene therapy. The multifunctional nanostructured carrier (i.e., nanobioconjugate) was based on the surface functionalization of magnetite with an organosilane molecule followed by a Polyether Amine (PEA) or a Polyethylene Glycol (PEG) surface spacer, and finally, a molecule containing a reducible disulfide linker. Simultaneously, a potent translocating, endosome escaping protein was co-immobilized on the surface to increase the efficiency of biologically active genetic material reaching the nuclei. The reducible disulfide linker is formed by direct conjugation of a thiolated 65 bp tag to the free thiol on the surface of the nanobioconjugate, which is complementary to a non-encoding sequence within the delivery vector. Once the nanobioconjugate reaches the intracellular space, the disulfide bond is reduced, and the cargo nucleotide delivered.

Magnetite was synthesized by chemical co-precipitation of the iron chlorides (II) and (III) in a sodium hydroxide 1M solution [28], [29]. Further details of synthesis can be consulted elsewhere [28], [29]. The nanoparticles were subsequently functionalized with the organosilane 3-Amino Propyl Triethoxy-Silane (APTES) to render free amine groups on the surface. The surface spacer PEA (Jeffamine 600) or PEG was conjugated followed by the 3-[(2-aminoethyl)dithio]propionic acid] (AEDP) containing a reducible disulfide bond [16]. The translocating molecule selected for this study was the Outer Membrane Protein A (OmpA) of *Escherichia coli* due to the recently reported penetrating and endosome escape abilities [28]. To investigate the role of the surface spacer, we prepared three different nanobioconjugates, the first one was produced by directly conjugating the AEDP to the surface of the silanized nanoparticles (OmpA-magnetite-AEDP) (Figure 1a), the second one contained PEG (OmpA-magnetite-PEG-AEDP) (Figure 1b) before AEDP and third one PEA (OmpA-magnetite-PEA-AEDP) (Figure 1c). For amine-amine conjugation, we used glutaraldehyde to form imine bonds, while for amine-carboxyl we used NHS/EDC to form amide bonds. The nanobioconjugates were characterized by FTIR, TGA, TEM, and DLS. The FTIR allowed us to confirm the characteristic bands of the polymer and OmpA. TGA analysis confirmed a conjugation efficiency of PEG of about 1.5% and of PEA of about 5% while for OmpA the efficiency approached 1.5%. TEM allowed us to confirm the typical morphology for the nanoparticles and an average diameter for individual particles of about 15-18 nm. Hemolysis and platelet aggregation exhibited values below 5%.

The delivery of the conjugated material was first tested *in vitro* with the aid of DTT and GSH as reducing agents. For the experiments, we conjugated BSA and Rhodamine and tracked the time evolution of the delivered molecules via Bradford and spectrofluorimetry, respectively. It appears that the highest release was obtained for the OmpA-magnetite-PEA-AEDP followed by OmpA-magnetite-PEG-AEDP and finally, OmpA-magnetite-AEDP. With these experiments, we decided to deliver to neuroblastome cells to evaluate endosomal escape. This was accomplished by labeling the nanobioconjugates with Rhodamine B (*OmpA-magnetite-PEA-AEDP-Rhodamine B*). Pearson's correlation coefficients were calculated to evaluate the colocalization of the nanobioconjugates with the endosomes (labeled with Lysotracker Green). The results suggest an endosomal escape of about 85%. Further evidence of escape and even cytosol distribution was provided by a calculated surface area coverage of about 60%.

Future experiments will be focused on the hybridization of a gene sequence for the expression of the fluorescent protein mCherry. The repair of missing nucleotides will be via homologous recombination using a Gibson Assembly kit. The obtained nanobioconjugate will be delivered to cells to evaluate transfection efficiencies. Detailed cytotoxicity experiments will be also required to assure a complete biocompatibility analysis.

[1] Abarca-Barriga, H. H., et. al. "Risk factors in genetic diseases ," *Acta Med Peru*, 2018.
[2] Rodríguez-Yunta, E. "Gene therapy and ethics principles," *Acta Bioeth.*, 2003.
[3] García, A. H., "Gene Therapy with Non-Viral Vectors", 2015.
[4] Austin-Ward, E. D., et. al. "Gene Therapy and Its Applications," *Rev. Med. Chil.*, 1998

[5] Y. Weng *et al.*, "Improved Nucleic Acid Therapy with Advanced Nanoscale Biotechnology," *Molecular Therapy - Nucleic Acids*, 2020

[6] Penn, S. G., et. al. "Nanoparticles for bioanalysis," *Curr. Opin. Chem. Biol.*, 2003

[7] Liu, T., *et al.*, "Calcium phosphate nanoparticles as a novel nonviral vector for efficient transfection of DNA in cancer gene therapy," *Cancer Biother. Radiopharm.*, 2005

[8] Mohammad, A., et. al. "Nonviral Vectors for Gene Therapy," *ScienceDirect*, 2015

[9] Lino, C. A., et. al. "Delivering crispr: A review of the challenges and approaches," *Drug Deliv.*, 2018

[10] Wang, Y., et. al. "Co-delivery of drugs and DNA from cationic core-shell nanoparticles self-assembled from a biodegradable copolymer," *Nat. Mater.*, 2006

[11] Plank, C., et. al. "Magnetofection: Enhancing and targeting gene delivery with superparamagnetic nanoparticles and magnetic fields," in *Journal of Liposome Research*, 2003

[12] Schillinger, U. *et al.*, "Advances in magnetofection - Magnetically guided nucleic acid delivery," *J. Magn. Magn. Mater.*, 2005

[13] Krötz, F., et. al. "Magnetofection Potentiates Gene Delivery to Cultured Endothelial Cells," *J. Vasc. Res.*, 2003

[14] Gersting, S. W. *et al.* "Gene delivery to respiratory epithelial cells by magnetofection," *J. Gene Med.*, 2004

[15] Wegscheid, M. L., et. al. "The art of attraction: Applications of multifunctional magnetic nanomaterials for malignant glioma," *Expert Opin. Drug Deliv.*, 2014

[16] Ramanujan, R. V., "Clinical Applications of Magnetic Nanomaterials," *Sch. Mater. Eng.*

[17] Kumar, C. S, et. al. "Magnetic nanomaterials for hyperthermia-based therapy and controlled drug delivery," *Adv. Drug Deliv. Rev.*, 2011

[18] Asín Pardo, L. "BIOMEDICAL APPLICATIONS OF MAGNETIC NANOPARTICLES: MAGNETIC HYPERTHERMIA IN DENDRITIC CELLS AND MAGNETOFECTION IN BRAIN CELLS."

[19] TARTAJ, P., *et al.*, *Biomedical Applications of Magnetic Nanoparticles*, 2007

[20] Kratz, H. *et al.*, "MPI Phantom Study with A High-Performing Multicore Tracer Made by Coprecipitation," *Nanomaterials*, 2019

[21] Chanana, M., et. al. "Fabrication of Colloidal Stable, Thermosensitive, and Biocompatible Magnetite Nanoparticles and Study of Their Reversible Agglomeration in Aqueous Milieu," *Chem. Mater.*, 2009

[22] Pan, B.F., et. al. "Dendrimer modified magnetite nanoparticles for protein immobilization," *J. Colloid Interface Sci.*, 2005

[23] Estrada de la Vega, A., et. al. "Aglomerados de nanopartículas basados en magnetita y CTAB," 2013

[24] N. Habibi, "Preparation of biocompatible magnetite-carboxymethyl cellulose nanocomposite: Characterization of nanocomposite by FTIR, XRD, FESEM and TEM," *Spectrochim. Acta - Part A Mol. Biomol. Spectrosc.*, 2014

[25] Xu, Z. P., et. al. "Inorganic nanoparticles as carriers for efficient cellular delivery," *Chem. Eng. Sci.*, 2006

[26] Fraile, M. C., "Study of the interactions between nanoparticles of noble metals and DNA," 2016

[27] Hsu, P. D., et. al. "Development and Applications of CRISPR-Cas9 for Genome Engineering," *Cell*, 2014.

[28] Lopez-Barbosa, N., *et al.*, "Magnetite–OmpA Nanobioconjugates as Cell-Penetrating Vehicles with Endosomal Escape Abilities," *ACS Biomater. Sci. Eng.*, 2019.

[29] Cuellar, M. *et al.*, "Novel BUF2-magnetite nanobioconjugates with cell-penetrating abilities," *Int. J. Nanomedicine*, 2018.

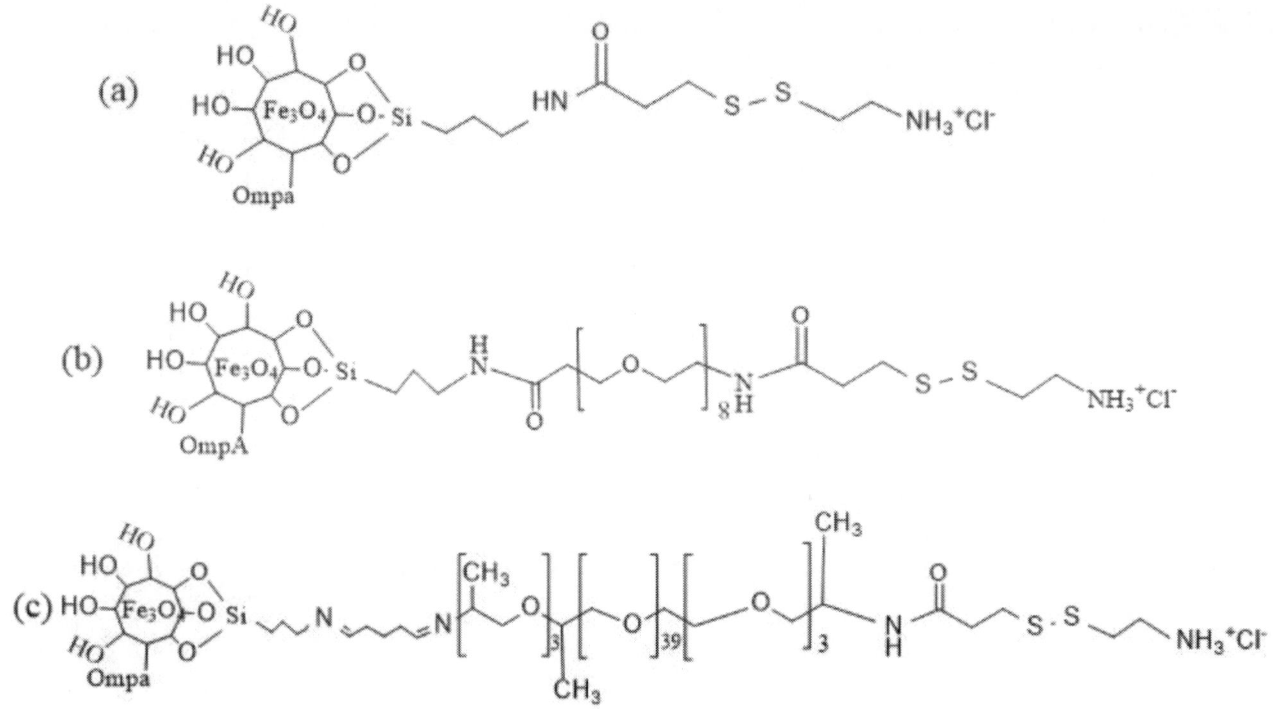

***Figure 1. Schematics of the prepared nanovehicles** (a) OmpA-magnetite-AEDP , (b) OmpA-magnetite-PEG-AEDP, (c) OmpA-magnetite-PEA-AEDP.*

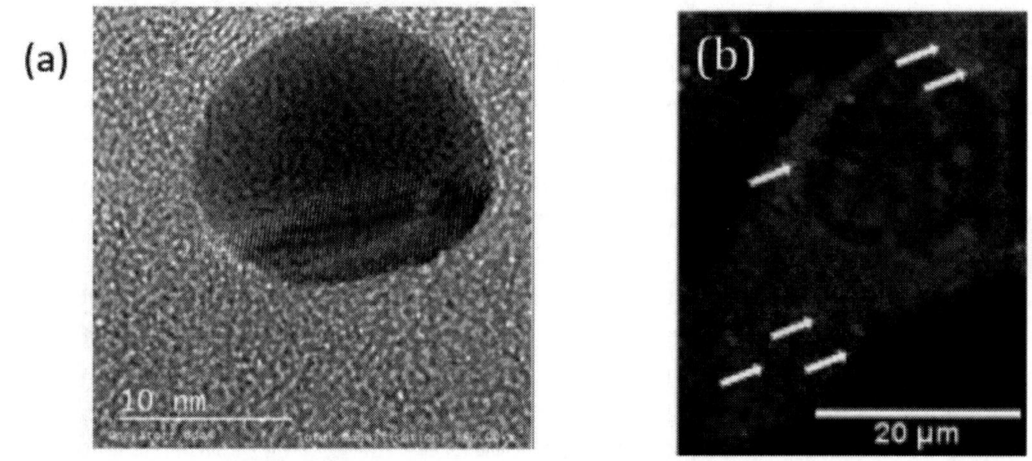

***Figure 2.** (a) TEM micrograph of magnetite. (b) Cell internalization of OmpA-magnetite-PEA-AEDP-Rhodamine B nanobioconjugates after 2 hours in neuroblastoma. Nuclei were stained with Hoechst and endosomes were labeled with lysotracker Green. White arrows point to regions of possible endosomal escape while yellow arrows point to regions of colocalization, which likely indicates endosomal entrapment.*

605614 Protein Nanoparticles for Effective Gene Delivery

Abstract:

Introduction: The largest hurdle to clinical application of gene therapies is the lack of a safe and efficient delivery system. To address this, we used electrohydrodynamic (EHD) jetting to develop a highly efficient, less cytotoxic plasmid DNA (pDNA) delivery system using protein nanoparticles (pNPs). pNPs are highly modular allowing numerous targeting ligands and various sized pDNA to be added without compromising ease of production. This system is viable in various cell lines and formulations, representing a promising platform for delivering various pDNA for applications in basic science and medicine.

Materials and Methods: *Synthesis and Characterization:* pNPs were fabricated using EHD (Figure 1A) then imaged using SEM and TEM (Figure 1B). The required weight ratio (N/P) between the PEI and pDNA was optimized and used in all pNP formulations (Figure 1C). Hydrodynamic diameter (D) and surface charge (Zeta Potential, ZP) of the pNPs were evaluated using a Zetasizer (Figure 1D, 1E). *In Vitro Testing:* Transfection was done by suspending pNPs in cell media for the duration of each timepoint. Cytotoxicity was measured using CellTiter-Glo (Figure 2A). pNP uptake was confirmed and quantified using confocal images and flow cytometry, respectively. The transfection efficiency (TE) of the system was measured with flow cytometry. TE measurements were taken at various loadings (weight percent of pDNA in the pNP) (Figure 2B, 2C). TE was measured for all GFP-positive pNP formulations with NLS addition, and the difference between corresponding pNPs was quantified. Timepoints were 24, 36, and 48 hrs. post transfection.

Results and Discussion: To generate modular pNPs, we utilized EHD jetting (Figure 1A) of human serum albumin (HSA) and transferrin, with polyethylenimine (PEI) 25kDa to electrostatically complex the pDNA and enable loading to protect against degradation. The versatility of the EHD jetting process lends itself to changing the base material and easily doping in peptides to serve as targeting ligands or NLS. Size, shape, and charge are critical to successful uptake and reducing pNP toxicity to the cells, which affects both transfection and viability of cells. The successful synthesis of spherical particles was confirmed using SEM (Figure 1B), and DLS and ZP were also quantified. Only pNP formulations that fit the criteria ($D < 200nm$, $ZP > -10mV$) were used going forward (Figure 1D, 1E).

After establishing a pNP delivery system that met the design criteria, we explored the effect of pDNA loading percentages and NLS peptide addition on system efficacy. The pNP delivery system exhibited both higher cell uptake and higher cell viability than PEI25k at comparable N/P ratios (Figure 2A). A potential reason for the increase in the observed cytotoxicity is the increase in total pNP delivered, resulting in an increase in PEI. This can be tested by delivering various concentrations of pNP at each loading weight percent. pNPs also showed higher or similar TEs at every tested formulations and N/P ratios while using empty pNPs, PBS as negative controls and PEI25k, lipofectamine as positive controls. The TE was further enhanced by addition of the NLS peptide, which was able to enhance TE in non-dividing cells, proving that the NLS facilitates active nuclear import of the pDNA, a large hurdle to the clinical application of non-viral systems. The higher relative cell viability combined with the ability to induce GFP expression shows a system which is more effective and less cytotoxic in all cell lines tested, for a large size pDNA as tested across a wide parameter space.

Conclusions: This work demonstrates that a new platform for pDNA delivery has been developed, demonstrating low cytotoxicity compared to current systems with a high TE and the ability to induce genetic changes in target cells. Further studies will include a study on the path of pDNA within the cell and quantifying transfection efficiencies, and PCR analysis to see genetic changes that result from delivered DNA. Unforeseen issues or sub-optimal efficacy can readily be adjusted due to the highly modular nature of this system, which will be integral to its adaptation for specific disease states.

Acknowledgements: This study was supported by the NIH (F31 HL149249).

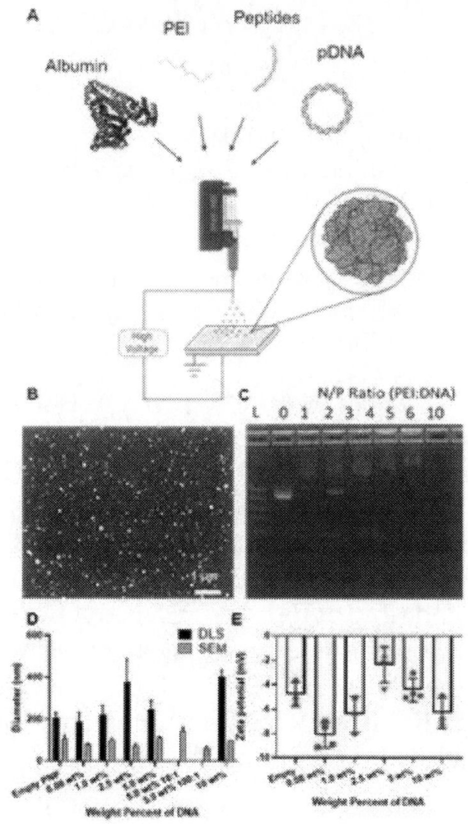

Figure 1. pNP Formation (A) EHD jetting process, which allows facile addition of various components to form pNPs. (B) Representative SEM micrograph of 5 wt% PNPs (scalebar 1 μm). (C) Gel complexation assay (D) The hydrodynamic (black, obtained using DLS) and dry (gray, obtained using SEM analysis) diameters of various pNP formulations. (E) Surface charges of various pNP formulations, obtained with Zetasizer.

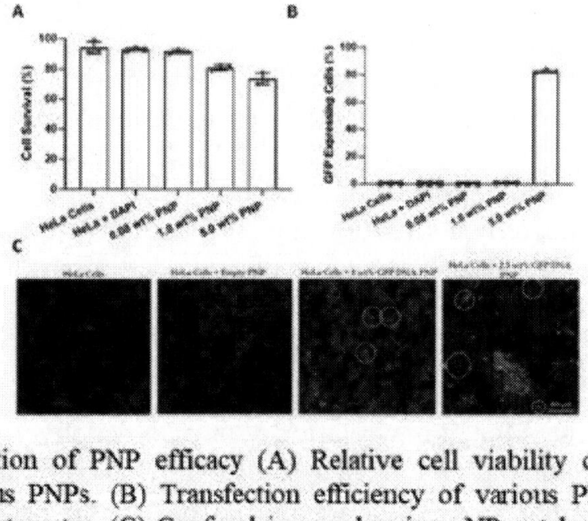

Figure 2. Characterization of PNP efficacy (A) Relative cell viability of HeLa cells at t=36h after transfection with various PNPs. (B) Transfection efficiency of various PNPs in HeLa cells at t=36h, quantified using flow cytometry. (C) Confocal image showing pNP uptake and some GFP expression in HeLa cells t=36h (scalebar 90 μm).

601743 Development and Toxicity Analysis of Carbon Nanoparticle Platforms for Gene Delivery into Plants

Abstract:

Plants will benefit from genetic engineering to feed our growing global population in the midst of climate change, to provide our most clinically-relevant drugs, and for our environmental remediation efforts. New generation genome engineering technologies, such as CRISPR/Cas, hold the potential to alter plant genes in a precise and targeted manner. The first step of any plant genetic engineering application is to deliver important biomolecular cargoes into the desired subcellular locations of plant cells. However, the limitations of traditional plant gene delivery methods, i.e. *Agrobacterium* and gene gun, prevent plant genetic engineering from reaching its full potential[1].

Given the lack of efficient biomolecule delivery vehicles to plants and potential impact of plant biotechnology on food security and sustainability, we developed nanoparticle-based platforms to efficiently deliver small interfering RNA[2,3] and plasmid DNA[4,5] cargoes into plant cells in a species-independent manner, which eliminates the use of gene gun and addresses the main delivery bottlenecks of plant genetic engineering. We show that pristine single walled carbon nanotubes (SWCNTs) can adsorb small interfering RNA molecules through *pi-pi* stacking for plant delivery, whereas a covalently modified and positively charged SWCNT platform is more efficient for large double-stranded DNA delivery to mature plant leaf cells.

This talk will first briefly summarize the development, characterization and biological verification of each nanoparticle platform for plant delivery applications. Next, I will present our recent results of CRISPR/Cas gene editing in plant leaves using the SWCNT DNA delivery platform. The talk will be concluded by discussing the several assays performed to demonstrate the effect of nanoparticles on plant health, including an RNA-sequencing analysis of nanoparticle-treated plant leaves. We believe the SWCNT delivery technologies that are discussed in this talk will have an immense impact on food security and sustainability by enabling the generation of improved plants that can rapidly reach the market in a cost-effective manner.

1. Cunningham, F.J., Goh, N., **Demirer, G.S.** et al. Nanoparticle-Mediated Delivery Towards Advancing Plant Genetic Engineering. *Trends in Biotechnology*(2018).
2. **Demirer, G.S.***, Zhang, H.*et al. DNA Nanostructures Coordinate Gene Silencing in Mature Plants. *PNAS* (2019).
3. **Demirer, G.S**. et al. Carbon Nanocarriers Deliver siRNA to Intact Plant Cells for Efficient Gene Knockdown. *bioRxiv* (2019).
4. **Demirer, G.S**. et al. High Aspect Ratio Nanomaterials Enable Delivery of Functional Genetic Material Without Transgenic DNA Integration in Mature Plants. *Nature Nanotechnology* (2019).
5. **Demirer, G.S**. et al. Carbon nanotube–mediated DNA delivery without transgene integration in intact plants. *Nature Protocols* (2019).

604910 Photo-Crosslinked Nanogel *Via* Nanoreactor for Therapeutic Proteins Delivery

Abstract:

Nanotechnology-based drug delivery systems use a variety of nanoparticle fabrication methods to formulate the different types of delivery formula. The nano-delivery systems provide the enhanced membrane permeability of the drugs because the nanoscale delivery formula allows the maximum surface area of a particle per unit of mass.

In this study, the polymeric nano-hydrogel (Nanogel) was formulated to deliver therapeutic proteins. The material biocompatibility and delivery efficiency were first taken into consideration to make an effective delivery carrier. Nanogel was synthesized by a process of confining hydrophilic biopolymer in a water-in-oil nanoreactor followed by photopolymerization and purification. This system enhanced the penetration of fluorescent-conjugated bovine serum albumin into a biological membrane indicating its suitability for transdermal drug delivery [1]. The modified nanogel was formulated by using fish gelatin methacrylate to enhance the drug diffusion at the target location without significant cytotoxicity [2]. For testing delivery efficiency and therapeutic effect, a model therapeutic protein was synthesized using a cell-free system described in the previous work [3]. The human skin protein was then encapsulated in the drug carrier and tested on the skin membrane model. Nanogel system showed great potential in transdermal drug delivery of protein therapeutics.

[1] *Macromol Res* 24 (12), 1115

[2] *Mar Drugs* 17 (4), 246

[3] *Method Protoc* 2 (3), 68

600493 Nanoalloy Catalysts for Boosting H_2 Production from Hcooh Decomposition: A First-Principles Study

Abstract:

The oxidation of small organic molecules has been widely studied owing to fundamental interest and relevance to fuel technology. Among these molecules, formic acid (HCOOH) has received much attention as a source of hydrogen because of low-toxicity, transportability, and easy-storability. HCOOH may decompose via dehydrogenation or dehydration, depending on the reaction conditions. It was found that the activity of the noble metal catalysts for the H_2 production via HCOOH decomposition was of the following order Pd > Rh > Pt > Au. Since the carbon monoxide (CO) generated from HCOOH dehydration can poison the Pt-based electrode in a fuel cell, a highly selective catalyst toward HCOOH dehydrogenation is required and such desired selectivity can be obtained by modifying the chemical and physical properties of surface by adopting the bimetallic Pd-M alloy catalysts, where M is transition metal. In this talk, I will deliver the underlying mechanism for the enhancement of H_2 productivity and selectivity in metal alloy catalysts. The followings are the main presenting topics. (1) We found that the H_2 production rate strongly depends on the Pd shell thickness in Ag-Pd core shell catalyst, in particular, the Pd thinnest shell (Pd monolayer) showed the highest activity. (2) In the bimetallic Pd/M catalyst, we discovered that HCOOH decomposition strongly relies on the variation of surface charge polarization (ligand effect) and lattice distance (strain effect), which are caused by the heteronuclear interactions between surface Pd and core M atoms. Especially, the contraction of bond distance between the surface Pd atoms and the increase of charge density in surface Pd atoms compared to the pure Pd (for example, Pd/Cu, Pd/Rh) are responsible for the enhancement of the selectivity to H_2 formation via HCOOH decomposition. (3) We employed Pd_3M-Pd core–shell models (M = Sc, Ti, V, and Cr) and discovered the Pd/Pd_3Sc and Pd/Pd_3V surfaces exhibit particularly enhanced selective H_2 production capability. Significant charge transfer from the subsurface Sc atoms to the surface Pd atoms and subsequent extremely low level of d band occupancy (<0.1) around the Sc atoms are identified as a key factor in deriving the modification of the Pd surface electronic structure. In contrast, the promotion by V is mainly attributed to the increase of the activation energy barrier of the CO producing reaction path as a result of the severe upshift in the transition state energy along with the relatively negligible modification of the initial state energy for the rate-determining step. Our theoretical calculation offers the fundamental mechanism of HCOOH decomposition on metal alloy catalysts and also provides physical and chemical intuition for the next generation bimetallic catalyst for hydrogen and fuel cell applications.

600729 Rational Design of CeO_2-Based Catalysts for Enhanced Water-Gas Shift Reaction

Abstract:

Water-gas shift reaction (WGSR) is important for chemical processes that use hydrogen resources. It has been widely used in fertilizer and petroleum refining industry. Nowadays, the WGS reaction is also aimed at preventing carbon monoxide (CO) poisoning which deactivates fuel cells. In the hydrogen flow supplied to the fuel cell, CO is contained, which is adsorbed on the anode to cause inactivation. Therefore, before the hydrogen is supplied to the anode, the WGS reaction process is performed to oxidize the contained CO to prevent poisoning. Supported metal catalysts are mainly used for the WGS reaction, and the support reducibility of these catalysts is widely known as an important factor that affects the activity. WGS reaction is catalyzed by redox and associative mechanisms, so the reducibility of catalyst plays an important role in the activity. Controlling the support reducibility can form a large amount of oxygen vacancies (O_v) on the catalyst surface. These O_v increase the lattice oxygen mobility, and help oxygen more easily participate in the oxidation reaction. Therefore, many studies to improve the support reducibility are actively underway. In this study, we prepared supported metal catalysts that controlled the reducibility of CeO_2 by metal-doping and morphology modification. First, we identified whether the catalyst structure was well formed from each approach. Subsequently, the synthesized catalyst was confirmed to have higher activity than un-modified catalysts and a reducibility test was conducted to prove that it was related to reducibility. Furthermore, we designed the catalyst using two strategies simultaneously to improve the reducibility described above.

600527 DFT Calculations and Machine Learning Approach to Predict Catalytic Properties of Nanoscale Electrocatalysts in Solution for Clean Fuel Generation

Abstract:

Innovative design of functional nanomaterials has been prolonged challenge to scientific community and industrial sector, especially regarding with structural stability on the thermodynamic and electrochemical aspects. It is due to the key functionality of the materials in wide range of applications, such as Li-air batteries, Photovoltaic devices, and fuel cells. Advanced experimental tools and computational theories attempted to deploy various candidates, but their performances in operations did not meet the practical level requirements.

For example, one of the major huddles toward to the solution is how to accurately and quickly explore the gigantic material space with varying the structure and composition of catalysts to identify promising catalysts and propose design principles for the performance optimization. Over the last several decades, first-principles calculations have been considerably developed for especially the objective through an efficient analysis of large-scale database. It is noteworthy that the approach enabled an identification of a key descriptor on the catalytic activity and the high-throughput screening of the best candidate beyond the conventional Pt. Unfortunately, most of the candidates were not applied to the real devices, due to the serious structural degradation issue via chemical or/and electrochemical reaction paths. It means that just electronic structure-level analysis for the catalytic activity may not be enough for designing high functional catalysts. Probably, the systematic and consistent incorporation of hybrid methodologies should be the paradigm. For example, establishment of big data, material informatics based on neural network or machine learning technique, and multi-scale simulations such as molecular dynamics (MD) or Monte Carlo (MC) simulations can be the relevant procedure.

This presentation introduces machine learning driven computational framework for high throughput screen of electrocatalysts for hydrogen fuel generation.

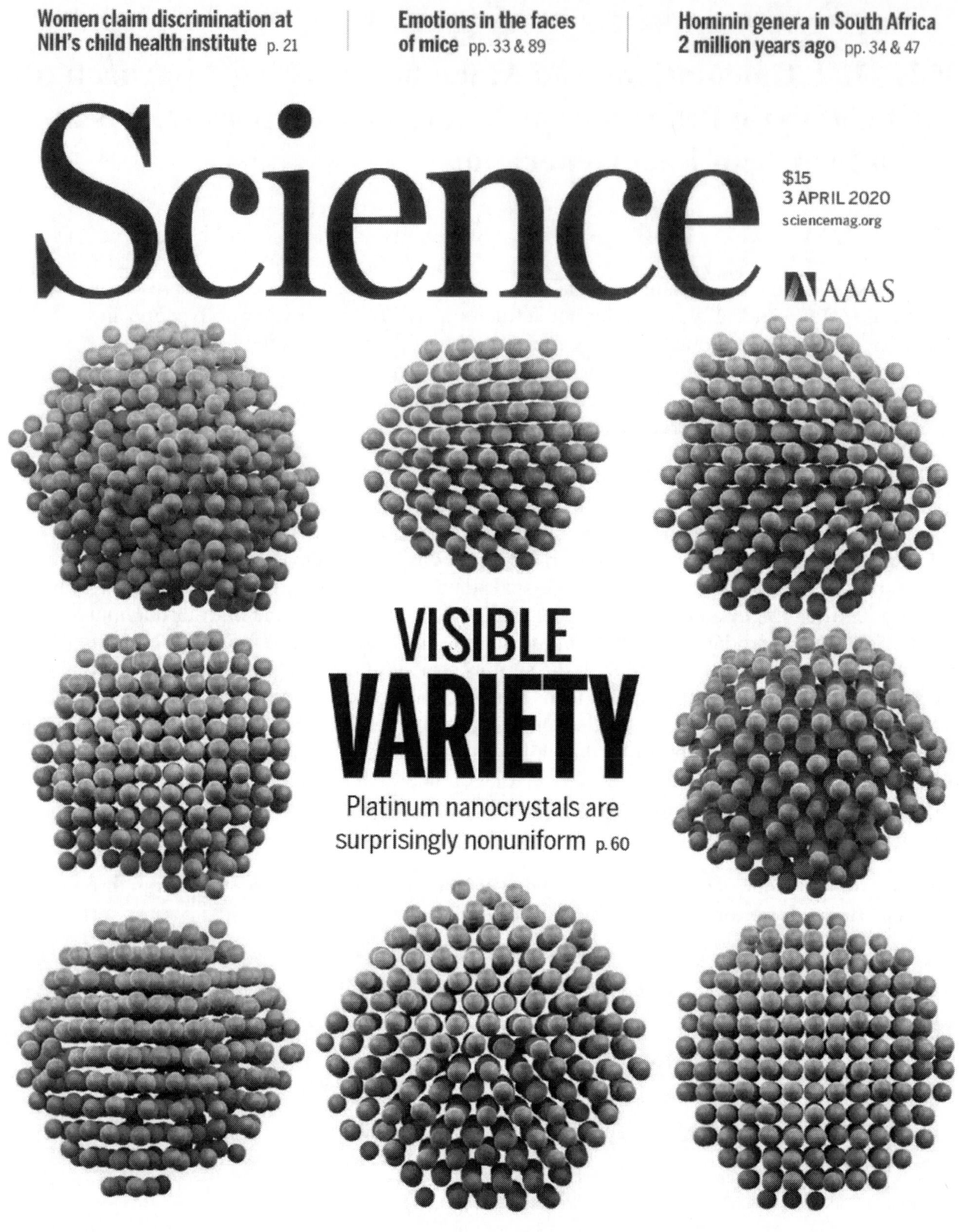
Women claim discrimination at NIH's child health institute p. 21
Emotions in the faces of mice pp. 33 & 89
Hominin genera in South Africa 2 million years ago pp. 34 & 47
Science
$15
3 APRIL 2020
sciencemag.org
AAAS
VISIBLE VARIETY
Platinum nanocrystals are surprisingly nonuniform p. 60

595901 Direct Access to Functional Porous Materials for Energy Conversion and Storage

Abstract:

Multifunctional and hierarchical porous materials have attracted much attention as host electrode materials for electrochemical energy conversion and storages. Our research group has developed powerful methods to control multiscale porous inorganic nanostructurs via simple "one-pot method" by employing blends of block copolymers and homopolymers. The new approach allows access to a high degree of control over pore structure and size, particle shape, particle size and chemical composition including metal oxides, metal nitride and conductive carbon. Multiscale porous materials have been employed as a multifunctional sulfur host, integrating the advantages of multiscale porous architectures with high performance electrocatalytic property to achieve high-power and long-life lithium-sulfur batteries. A new and intuitive strategy for tuning and enhancing the kinetic activity of Fe-N_4 sites was designed by controlling electro-withdrawing/donating properties of carbon plane. Fe-N_4 integrated mesoporous materials showed a high catalytic performance comparable to that of Pt/C in anion exchange membrane (AEM) fuel cells.

600504 Preparation of Multifunctional Carbon-Supported Irru Catalyst By Novel Acidic Polyol Procedure for Fuel Cell Application

Abstract:

Fuel cell systems have recently attracted attention as eco-friendly devices that can be applied to eco-friendly power trains for fuel cell electric vehicles and to combined heat and power generation, which is considered the important measures to the hydrogen society [1,2]. Among them, proton exchange membrane electrolyte membrane fuel cells (PEMFCs) have been mostly developed due to their possibility of ambient operation and high efficiency under the 100°C. In addition, the improvement of anion exchange membrane fuel cells (AEMFCs) has been remarkable along with the enhancement of the membrane performance during recent years [3]. These fuel cells are based on two basic electrochemical reactions such as hydrogen oxidation reaction (HOR) and the oxygen reduction reaction (ORR). To date, much research has been focused on searching the new ORR catalyst to increase the activity of the catalyst and to reduce the cost of the fuel cell systems. In the other hand, novel HOR catalysts is also demanded to be developed to address the issues from the Pt catalyst for PEMFC and AEMFC. For example, the Pt/C catalyst is very vulnerable under the fuel starvation situations that can occur during FCEV operation, which used the PEMFC system. In addition, in the case of AEMFC, it is known that the HOR activity of the Pt/C catalyst is drastically reduced.

In this presentation, as a universal HOR catalyst under acidic and alkaline, the carbon-supported IrRu alloy catalysts will be introduced. Recently, the IrRu alloy catalyst has been developed for replacing the Pt catalyst in the anode of the automotive PEMFC [4-6]. As a continuous study, the carbon-supported IrRu alloy catalysts with different ratios of Ir and Ru were prepared by polyol process under acidic condition ($pH < 7$) for the first time and investigated as the universal HOR catalysts under acidic and alkaline condition. The performance of the IrRu alloy catalysts in the single-cell level will be discussed.

[1] R.-H. Lin, Y.-Y. Zhao, B.-D. Wu, Int. H. Hydrogen Energy, in press (2020), doi.org/10.1016/j.ijhydene.2020.01.047

[2] The Fuel Cell Industry Review 2019 at www.FuelCellIndustryReview.com, E4Tech, (2019).

[3] D. R. Dekel, J. Power Sources, 375, 158 (2018).

[4] E. You, M. Min, S.-A. Jin, T. Kim, C. Pak, J. Electrochem. Soc. 165, F3094 (2018).

[5] S. W. Lee, B. H. Lee, T.-Y. Kim, C. Baik, M. S. Kim, G. S. Chai, C. Pak, Catal. Commun. 130, 105758 (2019).

[6] T.-Y. Kim, S. W. Lee, C. Pak, J. Ind. Eng. Chem. 85, 87 (2020).

598452 Electronic and Electrochemical Characteristics of Pt Monolayer on Graphene and Its Electrochemical Properties

Abstract:

Epitaxially grown Pt thin film on a graphene template is found to have a few monolayer thickness with structural stability comparable to Pt(111) surface and promising electrochemical activity. The newly synthesized thin film shows a quadrilateral-polygon structure of Pt in monolayer on graphene surface. In this study, we present a computational research on the unique architecture of the graphene-templated epitaxial Pt monolayer. Recently proposed strongly constrained and appropriately normed (SCAN) density function study (DFT) is employed to investigate the characteristics of the system. Electrochemical activities of the system are evaluated in terms of the free energy variation in oxygen reduction reaction (ORR). In architectures, Pt exhibits registry with the C-C bridge sites along the armchair and zigzag directions forming strong covalent bonds. Here, the details of the atomistic/electronic structures and binding energies are discussed. Interestingly, the ORR can occur on both Pt and graphene surfaces. This would provide a good strategy to protect metallic catalyst and to tune electronic structure of catalyst.

598810 The Counterintuitive Dependency of Bifunctional Effect on the Degree of Ordering in Alloy Electrocatalysts for Methanol Oxidation Reaction

Abstract:

Bifunctional effect is commonly regarded as the reason why bimetallic catalysts outperform pure-metal catalysts in processes that include CO oxidation. It has been shown that alloy catalysts shorten the distance between sites for CO_{ad} and sites for OH_{ad} into sub-nano range and are able to oxidize CO at much lower overpotential than non-alloyed catalysts with similar compositions. To see whether the highest intimacy between Pt and secondary metal in the ordered alloy could provide the highest activity, we modify the degree of ordering of Pt_3Sn nanocubes (NCs) while maintaining the exposed {100} facets, and test them in methanol oxidation reaction and CO oxidation. Strikingly, the 60% ordered Pt_3Sn NCs outperform the 95% ordered Pt_3Sn NCs in methanol oxidation by almost 5 times in converted charge density. 60% ordered Pt_3Sn NCs also oxidize CO at lower potential in CO stripping and are more effective in bulk CO oxidation than the 95% ordered Pt_3Sn NCs and Pt NCs. Moreover, we demonstrate the outstanding CO tolerance of less ordered Pt_3Sn NCs comparing to 95% ordered Pt_3Sn NCs by saturating the electrolyte with CO during methanol oxidation. We reveal that the Sn atoms at the surface of the 60% ordered Pt_3Sn are more prone to be oxidized under oxidative potential whereas the Sn atoms at the surface of the 95% ordered Pt_3Sn NCs are resistant to oxidation. We propose that the formation of SnO_x in the vicinity of Pt sites create higher proximity between OH_{ad} and CO_{ad} in less ordered Pt_3Sn NCs, making less ordered alloys more active for methanol oxidation reactions, in which CO oxidation plays a crucial role.

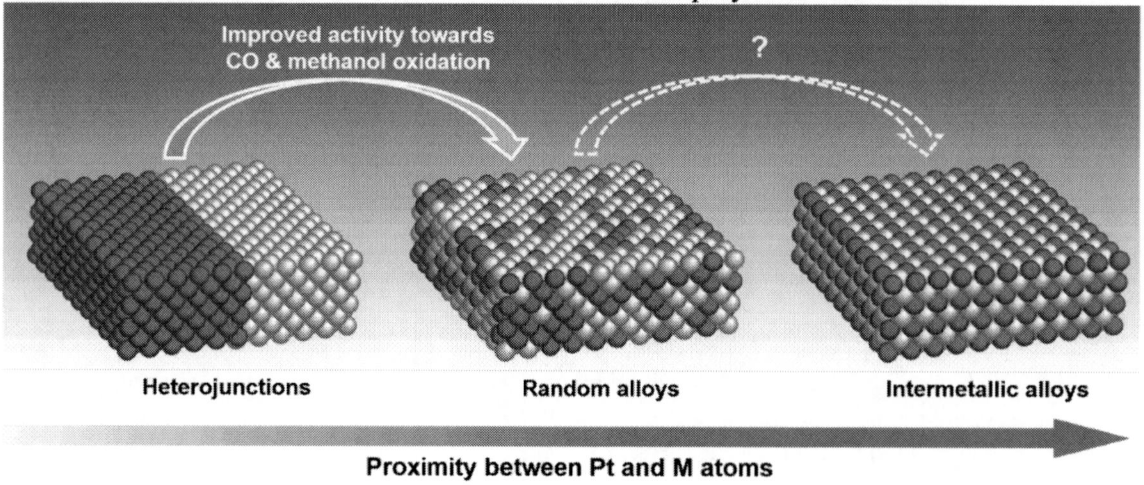

616170 Multi-Functional Biomaterials for Medical Applications

Abstract:

Tissue engineering is an interdisciplinary field incorporating concepts from engineering, biological sciences, and medicine with the goal of engineering biological substitutes to maintain, restore and promote normal tissue function. Polymer-based biomaterials have played an important role in the development of tissue constructs that mimic the structures and physical properties of the native tissues. However, there are still many challenges in micro-engineering biomaterials with tunable physical and biological properties for the development of fully functional tissue constructs. The combination of advanced biomaterials with micro- and nanoscale technologies have been shown to hold great potential to address the current challenges in tissue engineering. Our research focuses on developing biomimetic elastic biomaterials and nanocomposite hydrogels, with controlled architecture, and physical and biological properties, utilizing recombinant proteins. These advanced biomaterials are integrated with different micro- and nano-fabrication technologies to engineer biomimetic tissue constructs for engineering soft and elastic tissues. Furthermore, we have developed new chemistries to improve the adhesion of these biomaterials to the tissue surfaces and use them as multi-functional bioadhesives for sealing and repair of soft tissues such lung, heart, skin, and cornnea. In addition, our group has designed different nano delivery planforms which can be incorporated in these bioadhesives for gene and drug delivery applications. In this presentation, I will outline our recent work on the development of nanocomposite elastic hydrogels for tissue engineering along with their clinical applications as tissue adhesives and surgical sealants.

616171 A Journey from Energy Storage and Conversion to Nanomedicines

Abstract:

Living organisms are made from the basic elements such as C, O, N and P, harvesting and consuming energy to realize their reproduction. Driven by massive use of fossil energy (natural gas, coal and oil), industrial revolutions have built the civilization we are enjoying today. Clean energy and good quality of living, in this context, are two of the most essential components. In this talk, I will present my research in energy focusing on lithium-ion batteries, supercapacitors and fuel cells, as well in nanomedicines focusing delivery of protein therapeutics. From the angle of material design, a design principle towards better electrochemical devices, as well as a platform technology enabling effective delivery of therapeutics for treatments of alcohol intoxication, gout, central-nerve-system diseases, and COVID-19, will be discussed.

611832 Ultrathin Al-Doped ZnO Films Coated $Li_4Ti_5O_{12}$ As an Anode Material with Excellent Cycling Stability and Rate Capability for Lithium-Ion Batteries

Abstract:

Lithium titanate oxide ($Li_4Ti_5O_{12}$, LTO) is a lithium ion intercalated compound, which was identified as a promising anode material for LIBs with a theoretical capacity of 175 mAh g^{-1} within a typical potential range of 1.0 V–3.0 V. Extending the end voltage from 1.0 V to 0.01 V is an effective way to improve the electrochemical performance of LTO; the tetrahedral (8a) sites of LTO would be available for lithium ion storage to have an "extra" reversible capacity, when the end potential is lower than 0.6 V. However, this approach also leads to some other problems: (1) a lower end potential will cause a more intensive decomposition of electrolyte to form the solid electrolyte interface (SEI), which means that the initial irreversible capacity loss will increase, and (2) the change in lattice dimension during the lithium ion insertion/extraction will be increased and thus the structure stability of LTO will become worse. In this study, an Al-doped ZnO (AZO) film with various thicknesses and compositions were deposited on LTO particles by atomic layer deposition (ALD) method with different deposition temperatures and different ratios of the number of Al_2O_3 ALD to that of ZnO ALD. The electrochemical characterization results indicated that the coating of AZO significantly improved the electrochemical performance of LTO between 0.1 V and 3.0 V with a proper coating thickness. After 250 cycles of charge/discharge at a 1 C rate, the capacity of the optimum AZO-coated LTO sample decreased from 203 to 190 mAh g^{-1} with a capacity retention of ~94% at room temperature and from 224 to 216 mAh g^{-1} with a capacity retention of ~96% at 55 °C. The AZO coating layer with an appropriate thickness not only increased the capacity of LTO by enhancing conductivity, but also assisted the LTO electrodes to form a beneficial interface layer to protect the LTO from the continuous attack by harmful components in the electrolyte, especially under extreme conditions, such as high temperature and high current rates.

609488 Development of Novel Experimental Modules for Introducing Students to Nanoparticle Characterization Methods

Abstract:

Applications of nanoparticles in chemical and biomedical industries have been rapidly increasing in the recent years. The need to educate and prepare the students for this changing scene is increasing as well. Although new demands in skills have been met with efforts to introduce students to nanotechnology, most current efforts have focused on nanoparticle synthesis. Nanoparticle characterization, while equally important, has received far less attention. In this work, we have developed multiple nanoparticle characterization experiments to be performed as part of a newly developed "Nano and Biointerfaces" course. The aim is to introduce students of upper undergraduate level as well as graduate students in chemical engineering to various nanoparticle characterization techniques, including dynamic light scattering, scanning and transmission electron microscopy, BET surface area adsorption, and Fourier Transform InfraRed (FT-IR) Spectrometry. The underlying theory for each technique was introduced to students and students were asked to perform the experiments in coordination with the course teaching assistant or instrument technicians. To follow an inquiry-based approach, the properties of nanoparticles were not revealed to students prior to performing the experiments. The experimental modules were designed so that students can characterize nanoparticle composition, size, morphology, and surface area. The obtained results were presented in class by students followed by a discussion on the accuracy of their findings and comparison with the certificates of analysis for particles. Student laboratory memos and class presentations, a question in the final exam, and a student survey were the main methods of assessment. These educational activities can be improved by incorporating team-based assessments and evaluation of students' performance in teams.

600152 Radiation-Controlled Drug Release Formulation with Improved Therapeutic Index for Treatment of Head and Neck Cancer

Abstract:

Head and neck squamous cell carcinomas (HNSCC) are the 6th most common cancer in the United States predominantly affecting people over 65 years of age. Currently used modalities for treating HNSCCs include surgical resection, chemotherapy (CT) and radiotherapy (RT). Surgery is often not an option for the relatively older patient population due to exclusion criteria and post-operative decrease in the quality of life. For locally advanced HNSCC, combination CT-RT has been proven to be more effective than CT or RT alone and represents the current standard of care, but severe side-effects impacting a patient's quality of life persist. CT is associated with systemic toxicity causing grade 3 and grade 4 side effects post administration of chemotherapeutic drugs and poor localization causing secondary organ damage. Radiotherapy kills normal cells along with cancer cells along radiation beam tracks used for treatment and potentially lead to secondary cancers years post therapy. Intratumorally delivered chemotherapeutic drugs can potentially overcome some of these challenges by providing localized concomitant CT-RT. A calcium tungstate ($CaWO_4$)-based nanoparticle formulation (NPF) has been developed in our laboratory to facilitate controlled intratumoral CT-RT. Previous work has demonstrated the in vitro and in vivo efficacy of the formulation and demonstrated increased survival in mouse xenograft models as compared to primary radiation.

The formulation consists of a $CaWO_4$ core encapsulated with poly(ethylene glycol-block-lactic acid) co-polymer which is co-loaded with hydrophobic anti-cancer drugs that can be directly injected in solid HNSCC tumors. In the present work, we show that the rate of drug release can be controlled by varying the incident radiation doses which proceeds via a hydrolytic polymer degradation mechanism. The NPF has shown higher therapeutic index than clinically used modalities and is validated by histopathological analysis confirming lesser secondary organ damage. The in vivo efficacy of two different stereoisomers of Paclitaxel, a clinically used anti-cancer drug, vary with our NPF due to the different adsorption and release kinetics which sheds important information on how these interactions could be important towards designing nanomedicine formulations. A multi-compartment model has been developed based on pharmacokinetic release parameters obtained from in vivo experiments which predict the NPF can maintain the intratumoral drug concentration above the therapeutic limit for a significant time post single injection and can help predict clinical performances of the formulation.

These data collected for the NPF used as an alternative to systemic combination CT-RT is promising and can potentially be translated to the oncology clinic to improve treatment efficacy and provide a safer alternative to systemic CT and RT for head and neck cancers.

600261 Packaging and Delivering Enzymes By Amorphous Metal-Organic Frameworks

Abstract:

Crystalline metal-organic frameworks (MOFs) have been utilized to protect enzymes under harsh conditions[1]. Generally, there are two approaches to incorporate enzymes in MOFs: adsorption of enzymes in as-synthesized crystalline MOFs[2-5], and in-situ encapsulation of enzymes by crystalline MOFs[6-11]. However, the possibility of hosting molecules by amorphous MOFs with disordered and fuzzy structures has not been explored. We report the in-situ packaging of enzymes in amorphous MOFs (aMOFs) at ambient conditions. Mesopores in aMOFs as observed by cryo-electron tomography allow glucose oxidase (GOx) to display 20 times higher activity than that in crystalline MOFs. The delivery of GOx by aMOFs enables noninvasive measurement of glucose in a single living cell, which identifies cancer cells from normal cells. This capability shows great promise in cancer diagnosis and understanding tumor metabolism[12,13].

601751 Biothermal Analysis of Interactions between Delocalized Lipophilic Cation (DLC) and Composite Lipid Bilayer

Abstract:

Delocalized lipophilic cations exhibit a heightened biological effect as a result of interaction and passing through the phospholipid bilayer or the cell membrane. The interaction between delocalized lipophilic cations (DLC) and lipid bilayer results in changes in the physicochemical properties of the bilayer. In this work, the mechanism of interaction of a cyanine based DLC and a composite bilayer composed of charged lipids has been investigated. The nature of binding and internalization has been ascertained through thermo-analytic techniques such as nano DSC and nano ITC. The values of dissociation constant (Kd), Gibbs Free Energy (ΔG), partition coefficient (Pm), and no. of binding sites (n) from these techniques have been compared to determine if the binding affinity is enthalpy or entropy driven. It has been concluded that the lipophilicity of the cyanine molecule governs the number of the binding sites (n) required on the bilayer, which in-turn affects the enthalpy of the binding.

602224 Impact of Pore Size on Catalytic Behavior in Mesoporous Au@SiO_2 Core-Shell Nanoparticles

Abstract:

Mesoporous silica-encapsulated gold core-shell nanoparticles (Au@SiO_2 CSNPs) were synthesized via a bottom up procedure to facilitate the aerobic, solvent-free oxidation of benzyl alcohol. The pore size and morphology of the catalyst allows for the unique opportunity to investigate the impacts of pore size on mass transport and surface activity discretely. The nanoparticles exhibit an average pore diameter of 25.5 Å, a size which enhances selectivity via physical constraint to inhibit the formation of the smaller, desired product (i.e. benzaldehyde) relative to larger, undesired products (i.e. benzoic acid/benzyl benzoate). In addition to improving selectivity, the CSNPs also demonstrate an activity maintenance effect. Despite the mesoporous silica structure obscuring much of the gold active surface, GC-FID analysis revealed the observed yield was far higher than on bare supported nanoparticles of equivalent nanoparticle size and gold mass.

The pore size of CSNPs was controlled via additional calcination and etching in a weakly alkaline environment. Under these conditions, investigation using TEM revealed the etching process to have minimal impact on gold core diameter. GC-FID analysis revealed both lower activity and benzaldehyde selectivity on CSNPs with higher pore sizes, suggesting the mesopores inhibit both the formation of large, undesired products and competitive adsorption of reactant phenyl groups, which block reactions on the surface.

The pore length of CSNPs was controlled via a series of sequential silica condensations around the gold core, increasing the thickness of the encapsulating silica shell. Investigation using TEM revealed higher shell growth using multiple additions of silica precursor compared to a single addition of equivalent mass. Benzyl alcohol oxidation using CSNPs with higher silica shell thickness demonstrated lower catalytic activity as shell thickness increases. An activation energy study revealed the Thiele modulus transitions from purely rate limited to the transition region over shell thicknesses studied.

603028 Photothermal and Immunomodulatory Nanomaterials for Tissue Repair

Abstract:

Sutures, staples and tissue glues remain the primary means of tissue approximation and vessel ligation. Despite widespread use, conventional sutures do not immediately seal approximated tissue and are susceptible to bacterial leakage, wound re-opening and infection. Laser-activated tissue sealing is an alternative approach which conventionally employs light-absorbing chromophores and nanoparticles for converting near infrared (NIR) laser to heat. The local increase in temperature engenders interdigitation of sealant and tissue biomolecules, resulting in rapid tissue sealing. We developed laser-activated nanosealants (LANS) in which, gold nanorods (GNRs) or indocyanine green (ICG) dye are embedded within a biopolymer matrix (collagen, silk or elastin-like polypeptides). We also fabricated novel laser-activated tissue-integrating sutures (LATIS) that synergize the benefits of conventional suturing and laser sealing. These laser-activated approximation devices demonstrated higher efficacies for tissue biomechanical recovery and repair in a full-thickness, dorsal surgical incision model in immunocompetent mice compared to commercial sutures and cyanoacrylate skin glue. Localized delivery of modulators of tissue repair, including histamine and copper, further improved healed skin strength following laser sealing and application. LANS films loaded with antibacterial drugs were able to significantly lower MRSA loads at the wound compared to antibacterial sutures, indicating a multifunctional strategy that synergizes rapid sealing with combating surgical site infections. In addition to incisional wounds, histamine co-delivered with silk LANS films accelerated the closure of full thickness, splinted excisional wounds in immunocompetent BALB/c mice and genetically obese and diabetic db/db mice, resulting in faster closure than Tegaderm wound dressing. Histological and immunohistochemistry analyses showed LANS-histamine treatment reduced dermal gap, promoted angiogenesis (CD31+), myofibroblast-mediated wound contraction (aSMA+), and higher TGF-β1 expression which are hallmarks of improved healing outcomes. Growth factor proteins have been investigated in tissue repair but have demonstrated limited efficacies as monotherapies. In order to address this, we first generated and characterized fusion polypeptide growth factor nanoparticles (GFNPs) that demonstrated high stabilities in wound fluid. Combination treatments of LANS-histamine and GFNPs resulted in higher efficacies of wound closure and tissue repair compared to monotherapies acting alone. An investigation into the kinetics of wound closure indicated that temporally sequential delivery of LANS-histamine followed by GFNPs, coinciding with different stages of tissue repair, demonstrated the highest efficacies of wound closure and tissue repair. Our results indicate that laser sealing and approximation, together with delivery of immunomodulatory mediators, can lead to faster healing and tissue repair, thus reducing wound dehiscence, preventing wounds moving towards chronicity and lowering incidence of surgical site infections, all of which can have significant impact in the clinic.

604862 Scalable and Effective Dispersions of Boron Nitride Nanotubes By DNA in Alcohol/Water Mixtures

Abstract:

Boron nitride nanotubes (BNNTs) are one-dimensional cylindrical nanostructures with multifunctional properties that are unavailable in many existing nanomaterials for developing applications, such as high-performance composites, thermal management, and biomedical applications. Aqueous dispersions of BNNTs using various molecules, such as DNA and surfactants, were reported previously by ultrasonication. Although the debundling of BNNTs can be achieved, the formation of defects and cutting of BNNTs into shorter tubes are inevitable due to harsh ultrasonication. In addition, the mixed solvent strategy involving alcohol/water systems has been considered as efficient method to disperse nanotubes *via* a mild bath sonication, which preserves intrinsic properties of BNNTs. In this work, we investigated the effect of various alcohol/water mixed solvents including methanol, ethanol, and isopropanol on stabilizing BNNTs using DNA. The solvation behaviour of BNNTs in a range of alcohol/water mixtures was investigated by molecular dynamics simulations. In addition, length distributions of DNA-wrapped BNNTs were determined by SEM imaging and rheology to better estimate the apparent length distributions of nanotubes. This study addresses the scalable and efficient dispersions of BNNTs which is an important step for producing nanotube-based materials such as films and fibers with enhanced properties.

605195 Laser Induced Morphology Change in Copper Sulphide Nanoparticles.

Abstract:

Copper Sulphide nanoparticles (CuS_0) were synthesized by reacting copper acetate with thiourea in oil-in-water microemulsions under high temperature conditions (65°C). The prepared nanostructures were characterized using UV-visible spectroscopy, SEM, TEM, Raman, EDAX for their size, morphology, behavior and structure. These copper sulphide nanoparticles stabilized with Pluronic F-127 were found to be spherical in shape possessing λ_{max} at 922nm. The average diameter of the nanoparticles determined by TEM was 18nm. The prepared particles showed broad absorption spectrum in NIR region. The prepared particles were irradiated by using laser of 808nm (NIR region) for 10min(CuS_{10}), 20min(CuS_{20}) and 30min(CuS_{30}) to investigate its effect on morphology and behavior. The changes had been studied by TEM, SEM, Raman and by observing the shifts in UV absorbance peaks. Upon exposure to laser radiation, the monodispersed nanoparticles were found to assemble like flowers, small rods and long extended rods. Photo thermal efficiency keeps on decreasing with irradiation time. The nanoparticles were found to be amorphous initially which became crystalline, upon irradiation with laser light after 30 min. The possible criteria or mechanism involved in the formation of different nanostructures has also been discussed.

605198 Antibody Drug Nanoparticle Induces Synergistic Therapeutic Outcome in Breast Cancer

Abstract:

We explore a combination therapeutic approach of chemo and immunotherapy in exploiting the specificity of antibody conjugated drug nanoparticles for breast cancer therapy. The goal of this work is to engineer antibody-drug nanoparticles (ADNs) comprising of pure anti-cancer drug nanorods (NRs) in the core with a therapeutic monoclonal antibody on the surface of the nanoparticles for specific targeting and synergistic treatments of cancer cells. The ADNs were designed by first synthesizing ~95 nm diameter × ~500 nm long paclitaxel (PTX) NRs using nano precipitation method (Figure 1a,1b). The surface of PTXNRs were functionalized at 2'OH nucleophilic site using carbonyl diimidazole and conjugated to Trastuzumab (TTZ) through the lysine group interaction forming PTXNR-TTZ conjugates. The optimum conjugation efficiency was >95% with the combination of 5 mg/ml PTXNR and 0.5 mg/ml TTZ. The drug NRs along with the antibody conjugated PTXNR-TTZ showed substantial colloidal stability. *In vitro* cytotoxicity analysis showed PTXNR-TTZ inhibited the HER2-positive breast cancer cells more efficiently than the individual treatment with PTXNR and TTZ . PTXNR-TTZ inhibited >80% of the HER2-positive breast cancer cells whereas in case of HER2-negative breast cancer the inhibition was roughly 50%. In both the cell lines conventional Paclitaxel solution showed resistance with increasing dosage. Whereas, PTXNR monotherapy or PTXNR-TTZ combination treatment had increased toxicity with increasing dosage. PTXNR-TTZ showed a synergistic effect on HER2-positive breast carcinoma. Cell cycle analysis showed PTXNR-TTZ arrested more than 80% of the cancer cells in G2/M phase (Figure 1c). In western blot analysis protein expressions show that the combination treatment with PTXNR-TTZ drug nanoparticle inducing prolonged cell cycle arrest in G2/M phase resulted in apoptosis in HER2-positive breast cancer cells in the intrinsic caspase dependent pathway (Figure 1d).

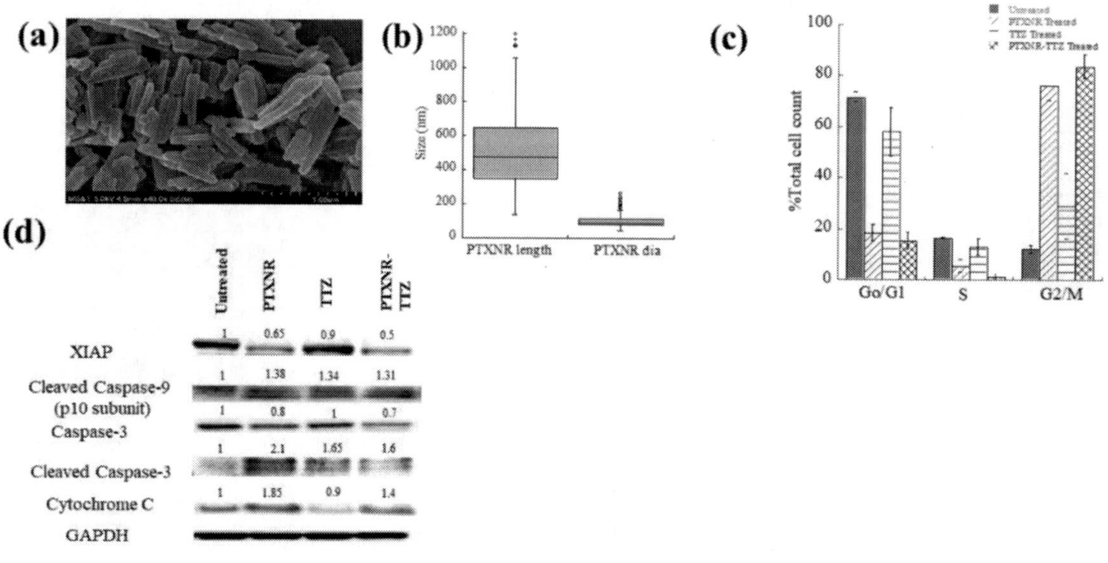

Figure 1. Characterization of PTXNR nanoparticles and G2-M cell cycle arrest by combination drug PTXNR-TTZ. (a) scanning electron microscopy (SEM) image of PTXNR nanoparticles (b) percentile plot of the PTXNR particle length and diameter (c) cell cycle analysis after 72 hr in BT-474 breast cancer cells (d) western blot analysis in BT-474 breast cancer cells.

606981 Gold Nanoparticle Formation in Water-in-Oil Microemulsions:Experiment and Simulation

Abstract:

Self-assembled, surfactant templated microreactors, such as water in oil (W/O) microemulsions, have been reported as a suitable route for synthesis of controlled size nanoparticles. These microemulsions are thermodynamically stable, nanoscale droplets of water dispersed in a continuous oil phase. The main advantage of microemulsion is that the drop size can be easily tuned to get the desired particle size. However, the mechanism of formation of nanoparticles in microemulsion is still open for discussion. In this work, gold nanoparticles (GNPs) were synthesized via W/O microemulsion route. As the molar ratio of water to isooctane (R) was increased from 2.5 to 5.0 to 7.5, its corresponding drop diameter increased from 2.7 to 5.0 to 7.3 nm, respectively. The mean diameter of GNPs simultaneously increased from 6.5 to 11.3 to 15.6 nm, for correspondingly increasing values of R, i.e., 2.5 to 5.0 to 7.5, respectively. Thus, for all values of R, the average diameter of the GNPs was significantly larger than the initial droplet size. It is clear that, existing kinetic Monte Carlo (kMC) simulation underpredicts experimental results. However, only on including coalescence of drops during particle coagulation event, the current kMC simulation agrees well with the experimental results. We also find that, coagulation efficiency (β_p) increases with R, implying larger drops coalesce more easily. Hence, a combined study of simulation and experiment enumerates the dynamics of size evolution of nanoparticles and the events involved in the formation of nanoparticles in W/O microemulsions (Figure).

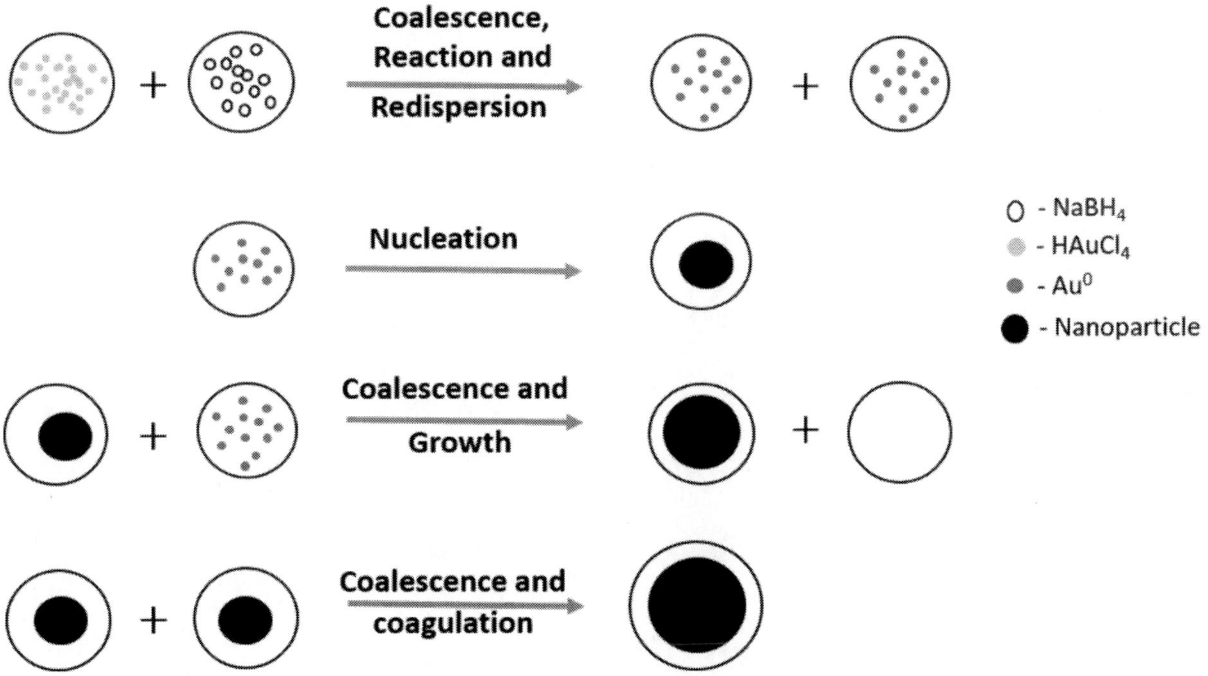

607711 Characterization and Application of Laboratory Grade Nanoscale Egyptian Blue

Abstract:

Calcium copper silicate ($CaCuSi_4O_{10}$), known as Egyptian Blue (EB), is a pigment that exhibits a strong near-IR (910 nm) fluorescent emission when exposed to a strong visible light source. This property makes it a novel candidate for use as a basis for nanomaterial-based sensors in surface water, as its IR signal is very strong in comparison with current commonly used IR reporters. Also, IR signals penetrate biological samples better, and are scattered less than visible range signals. A sample of EB was synthesized by a solid-state reaction outlined by Johnson-McDaniel and Salguero. In order to reduce the particle size to nanoscale, exfoliated EB powder and exfoliated EB powder covered with cetyltrimethylammonium bromide were prepared. The powders were adhered to glass discs in conjunction with (3-aminopropyl)triethoxysilane to develop lens coatings that fluoresce in the near-IR spectrum. Dynamic light scattering was performed to obtain information on particle sizes. Phase shifting interferometry was performed to characterize the surface coatings of the glass discs and to obtain the size distribution of the adhered nanoparticles. The performance of the lens coatings was characterized by photoluminescence spectroscopy.

608560 Near-Infrared Optical Detection of Dopamine Using Xeno Nucleic Acid (XNA) Sensors

Abstract:

Single-walled carbon nanotubes (SWCNTs) emit near-infrared fluorescence that is beneficial for in vitro and in vivo optical sensing. State-of-the-art SWCNT sensors are functionalised with single-stranded DNA (ssDNA) sequences that control the dispersivity, brightness, biocompatibility, and selectivity of the sensor. However, the fluorescence of the ssDNA-functionalized nanotube sensors is sensitive to changes in the local ion concentration. This sensitivity is problematic not only for measurements in complex biofluids, but also for sensing neurotransmitters, whose concentrations are strongly coupled to spatial and temporal ion fluctuations. The fluctuating ion concentrations can therefore interfere with the accurate optical detection of neurotransmitters, thereby limiting the use of ssDNA-SWCNT sensors for detecting and treating neurological diseases.

We present an optical SWCNT dopamine sensor with improved stability towards cation-induced fluorescence changes. The synthetic biology approach presented in this work serves as a complementary means for enhancing nanotube optoelectronic behaviour, unlocking previously unexplored possibilities for developing nano-bioengineered sensors with augmented capabilities.

608730 Machine-Learning Driven Potential Energy Surface for Nanoparticles Alloy System

Abstract:

In developing machine-learning potential energy surface for nanoparticles, it is difficult to map the training data of the structure. Conventional input symmetry functions of atomic distance and angle is restricted in bulk system. However, the structures of nanoparticles are dynamic which makes the machine-learned potential predict the energies and forces. Therefore, it is of importance to account structures that are off from the crystalline. Here, we obtain the structures from the real TEM images, understand the effect of input structures for the machine-learning potentials and extend the methodology to alloy system to characterize the stability of the nanoparticles. This will provide the efficient way to investigate the nanoparticle growth mechanism.

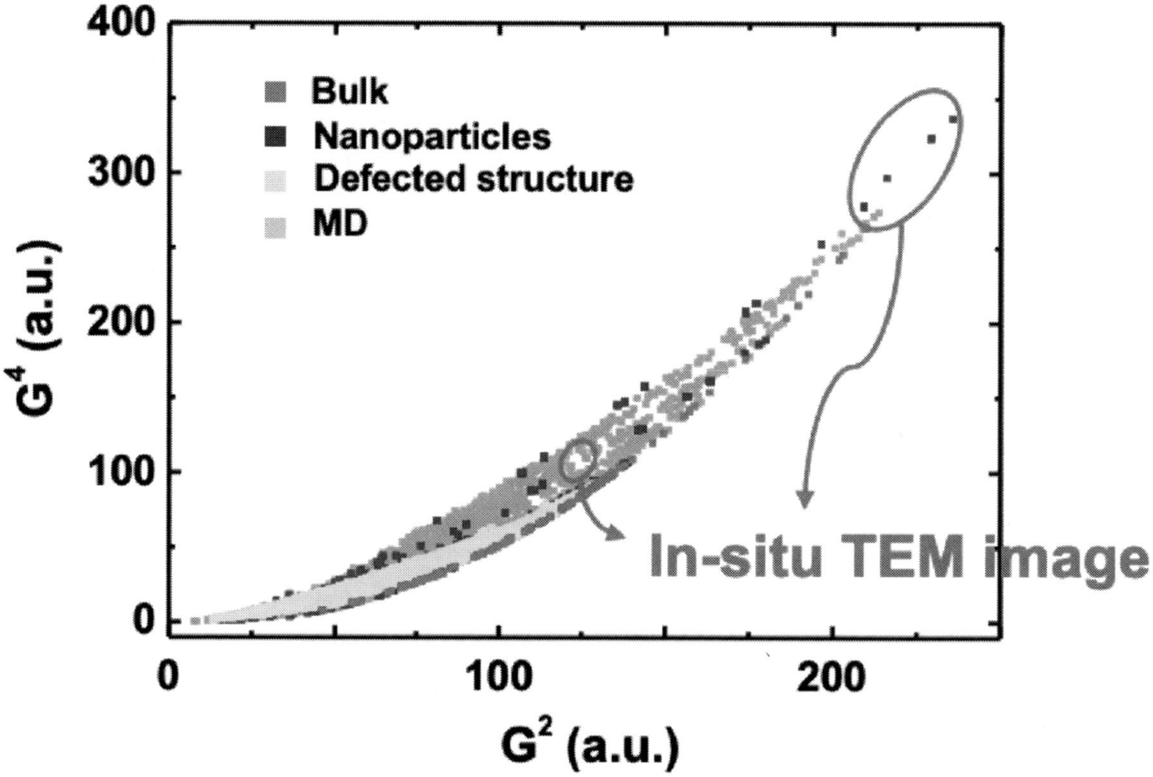

609174 Aqueous Dispersion and Characterization of Glycopolymers-Wrapped Carbon Nanotubes

Abstract:

Single-wall carbon nanotubes (SWCNTs) have gained tremendous attention for both material and biomedical research and applications, such as bioimaging, biosensor, and drug delivery applications. Dispersion of SWCNTs with biopolymers in aqueous environments facilitates their potential biological and biomedical applications. In this work, we investigated a small library of synthetic glycopolymers with monosaccharide and disaccharide groups for dispersion of SWCNTs *via* non-covalent complexation in aqueous condition. Glycopolymers exhibited the effective stabilization of SWCNTs in water through the formation of wrapping conformation on the nanotube surface, which also depends on the carbohydrate density and polymer chain length. This work demonstrates the synergistic effects of the identity of carbohydrate pendant groups and polymer chain length of glycopolymers on stabilizing SWCNTs in water, which has not been achieved previously.

609296 Improving the Pattern Transfer Process Using AAO Templates on Hydrophobic Substrates

Abstract:

Periodic nanopore array patterning of surfaces has received enormous attention in recent years due to its unique and enhanced optical, electrical and transport properties. Nanopatterning surfaces by template assisted pattern transfer technique provides a cost- effective and large throughput approach than e-beam lithography and nanoimprinting lithography techniques. Anodized Aluminum Oxide is a popular nanomaterial used as a mask in template assisted pattern transfer techniques for fabricating nanopores, nanowires, nanodots etc. It is self-organized nanostructured material with high aspect ratio structure, and hexagonally ordered straight nanopore channels. In template assisted fabrication using AAO, there are two ways of contacting the AAO mask on the substrate of interest: the AAO can be directly grown on evaporated/sputtered Aluminum layer on the substrate or by transferring a free standing AAO membrane grown separately on a thick aluminum sheet to the substrate of interest. The second approach is usually preferred due to its high pore density and template quality. Through this approach, nanopore arrays have been fabricated on various hydrophilic substrates like Si, GaAs, InP, GaN and ZnTe etc. However, when this process is performed on hydrophobic polymers with low surface energy, the template transfer is limited to smaller areas due to poor contact between the AAO membrane and the substrate. Apart from this, other factors like improper removal of the polymer support layer due to polymer solution wetting the pore channels, and incomplete barrier layer removal due to non-uniform barrier layer thickness across the AAO template limits the pattern transfer area to few hundreds of μm^2. In this work, the factors affecting the pattern transfer process on Teflon using AAO templates were identified, and a modified processing technique is carried out to improve the pattern transfer process. By applying the modified procedure, we were able to achieve pattern transfer over a large areas of $inch^2$ on Teflon.

609824 Fabrication of Polyelectrolyte-Surfactant Nanoparticles in T-Sensor Type Microfluidic Chips

Abstract:

This work is focused on the research in interactions of polymers and surfactants in a non-equilibrium microfluidic confinement. The main factor of our research is that microfluidic convection-diffusion conditions bring up multiple additional factors that affect complexation in reacting polymer and surfactant flows. These factors allow proceeding synthesis of non-equilibrium polymer-surfactant complexes that are impossible to produce in macroscopic conditions. Factors that are represented as dimensionless parameters come as convenient tools to predict locations of reaction fronts in a microchannel control the phase behavior of confined polymer-surfactant solutions and sizes of associates.

Polymer-surfactant nanoparticles were synthesized from Polydiallyldimethylammonium chloride (PDADMAC) and Sodium dodecyl sulfate (SDS).

For macroscopic experiments, PDADMAC and SDS solutions were mixed in different ratios.

For microfluidic experiments, PDADMAC and SDS solutions were infused into microfluidic devices by syringe pumps. For images recording were used Levenhuk D320 optical microscope at 10X magnification. Microfluidic chips were fabricated by standard photolithography technique using Sylgard 184 polydimethylsiloxane. The selected design for microchips was the T-sensor (3 inputs for solutions of polymer and surfactant and a solvent). The width of microchannels is 200 μm.

PDADMAC-SDS systems were initially characterized in macroscopic conditions. With increasing level of surfactant concentration, polymer-surfactant nanoparticles form in solutions. The solutions become turbid and this turbidity increases with more added surfactant. The nanoparticles are represented by aggregates of macromolecules with bound surfactant ions. According to dynamic light scattering measurements, the hydrodynamic radius of such complexes is approximately 100-200 nm. When surfactant-polymer concentration ratio reaches 0.4-0.5, precipitates form in solutions and turbidity decreases. Such a behavior is the result of formation of insoluble associates through neutralization of charges on PDADMAC macroion by additives of oppositely charged SDS ions and aggregation of the resulting hydrophobic complexes. Precipitation occurs when the residual charges on a macromolecular chain are not sufficient to keep macromolecules in aqueous solution.

The results of macroscopic characterization contributed to planning microfluidic experiments. Polymer-surfactant nanoparticles were synthesized in microchannels by radial diffusive mixing of parallel laminar flows of polymer and surfactant solutions separated by a solvent flow. The solutions of complexes with surfactant-to-polymer ratio Z=0.1, 0,3, and 0,5 were also mixed with surfactant solutions in the experiments analogous to those with the polymer flows.

Microfluidic studies of PDADMAC and SDS complexation revealed considerable differences between the results of experiments performed in macroscopic conditions and in microfluidic confinement. In opposite of macroscopic experiments, microfluidic processes involving interactions of polymers and surfactants occur in non-equilibrium conditions. The reaction of complexation takes place after mutual diffusion of PDADMAC and SDS after junction in the flow of solvent. Accumulation of the reaction product in the narrow region of the reaction front results in precipitation even if the respective polymer-surfactant system is homogeneous. As the result, the straight line of precipitate forms in microchannels. The location of this line depends on diffusion coefficients of the reagents.

Dynamic light scattering studies of PDADMAC-SDS solutions taken at the microchannel output demonstrate that the hydrodynamic radii of associates vary in the range of 100-200 nm that is in agreement with macroscopic experiments. The sizes of aggregates can be reproducibly controlled in this range by varying the flowrates of the reagents and the solvent.

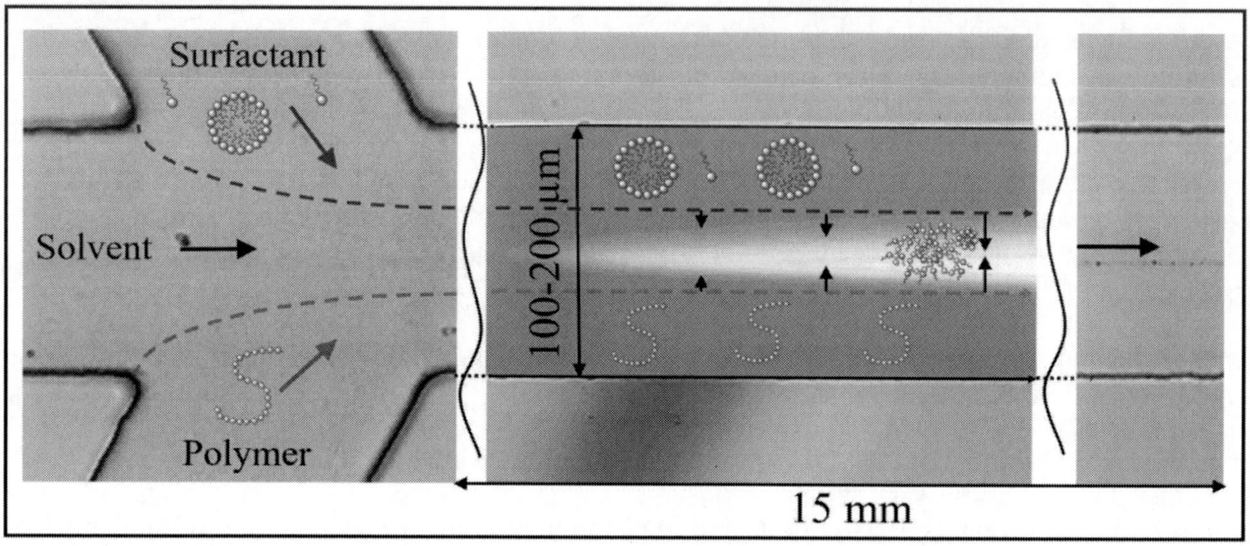

610420 Transition Metal Phthalocyanines Catalyst on Nano Scale Photothermal Support for Solar Mediated Methane to Methanol Conversion

Abstract:

The goal of this work is to find a suitable catalyst based on a transition metal phthalocyanines (TMPc) catalyst on photothermal support for the conversion of methane to methanol using solar energy. The conversion of greenhouse gas methane into methanol is important as methanol can be used as liquid feedstock and fuel which can be easily transported and handled. The conversion process usually requires high temperatures or pressures to ensure the partial oxidation of methane due to the very high tetrahedral C-H bond dissociation energy of methane (~4.51 eV). However, natural monooxygenase enzyme containing metal-oxo species can convert methane into methanol even at room temperature. These motivated the search of catalyst based on single-atom active sites of metal phthalocyanines (analogous to metal-oxo species of monooxygenase enzyme) deposited on the catalytic supports for methane to methanol conversion.

Here we evaluated graphene(G) and titanium nitride(TiN) nanoparticles as photothermal support for TMPc catalysts. G and TiN were chosen as substrates due to their high electron mobility, strong light absorption and photothermal properties. Moreover, titanium nitride nanomaterials are refractory plasmonic materials hence can remain stable at high reaction temperatures. Combined theoretical and experimental studies are performed to understand the electronic and material properties of G-TMPc and TiN-TMPC systems and relate that to their efficiency towards C-H bond activation. The TMPc's chosen for study are Cobalt-Pc(CoPc) and Copper-Pc(CuPc) as theoretical calculations show these materials to have low, and sometimes negative energy barriers for the formation of Oxides on their metal cores when supported.[1] The calculations have also suggested that the oxidized metal cores have favorable energies for cleaving the C-H bond of methane. Temperature dependent diffusion reflectance FTIR spectroscopy study suggested that both G supported CoPc and TiN supported CoPc systems begin to oxidize at lower temperature than their CuPc counterparts. This agrees with theoretical findings showing a higher energy barrier in CuPc Systems (2.1 eV) in comparison to that of CoPc (0 .8 eV). It has been found that cobalt-oxo species and copper-oxo species can be formed on TiN support at room temperature and 150^0C respectively. The temperature of the TiN support could be increased to 150^0C using two sun intensity. Most importantly, the C-H bond breaking energy for this copper oxo-species on titanium nitride support is favorable for room temperature C-H bond breaking. The electronic interaction among the support and the TMPc molecule is an important descriptor which decides their efficiency towards both metal oxo species formation and CH bond breaking. It has been found from photon emissive electron microscopy, (PEEM) studies of graphene/CoPc molecule, the graphene acts as an electron donor. The charge transfer from graphene to metal oxo-species has favorable effect toward metal-oxo species formation energy but has less effect on C–H bond activation energy.[1]

1. Filonowich, D., *et al.*, *The Journal of Physical Chemistry C* (2020) **124** (8), 4502

610504 Structure and Dynamic Property of Electric Double Layer in Ionic-Liquid-Gated Transistor: The Anatomy of Enhanced Gating Performance

Abstract:

Ionic-liquid-gated electric double layer transistors (IL-gated EDLTs), in which the ultrahigh carrier density at semiconductor surface could be achieved by EDL formed by ILs, are attracting growing research interests. Since the interfacial properties of EDL greatly affect the performance of IL-gated EDLTs, understanding the underlying mechanisms is of great significance for designing EDLTs with prominent performance.

In this poster, using molecular dynamics (MD) simulations, the interfacial structures and dynamic properties of EDL formed by different ILs were analyzed. Our previous work shows that during gating process, the more homogeneous electric field-induced charge distribution at semiconductor surface would result in the higher electrical conductance of semiconductor.[1] By comparing the ionic liquids with different cation types, we found that the type of cation functional group has a great impact on the homogeneity of charge distribution at semiconductor surface, and the conductivity of semiconductor is finally determined by both the homogeneity of charge distribution and the capacitance. Besides the static properties, we found that by adding organic solvent, both the response speed of IL at semiconductor surface and the homogeneity of semiconductor surface charge distribution could be significantly improved, which means that the dynamic and static gating performance of IL-gated EDLTs could be improved simultaneously.[2] Moreover, the microscopic properties obtained by MD simulations were successfully bridged with macroscopic characteristics of IL-gated EDLTs by resistant network method. Overall, our findings provide both new concepts and modeling techniques for IL-gated EDLTs.

[1] Zhao W, Bi S, Balke N, Rack PD, Ward TZ, Kalinin SV, Dai S, Feng G. Understanding Electric Double-Layer Gating Based on Ionic Liquids: from Nanoscale to Macroscale. *ACS Applied Materials & Interfaces.* 2018; 10:43211-43218.

[2] Zhao W, Bi S, Zhang C, Rack PD, Feng G. Adding Solvent into Ionic-Liquid-Gated Transistor: The Anatomy of Enhanced Gating Performance. *Submitted.*

610789 Adsorption Isotherms of Carbon-Framed Magnetic Iron Oxide Nanoparticle Adsorbents in Selenium Contaminated Water

Abstract:

Demand for potable and irrigation water is increasing as the world population grows, however, contaminants present in water pose a safety risk. Selenium is naturally present in bodies of water but is highly toxic if concentrations exceed regulatory guidelines. Removal of selenium is primarily done through the use of adsorbents. Magnetic iron oxide nanoparticle adsorbents suspended within a tire-derived carbon framework (C-MNA) are compared with activated carbon adsorbents and their adsorption isotherms are modelled using a Freundlich fit.

610897 A Corona Phase Hamiltonian for Cylindrical Nanoparticle-Polymer Interactions

Abstract:

Nanoparticle surfaces, such as cylindrical nanowires and carbon nanotubes, are commonly decorated with adsorbed polymer corona phases to impart solution stabilization and to control molecular interactions. While the specific configuration of this corona phase is critical to interfacial interactions, the prediction of its most basic properties remains an unresolved problem in polymer physics. In this work, we construct the Hamiltonian describing the adsorption of an otherwise linear polymer to the surface of a cylindrical nanorod in the form of an integral equation summing up the energetic contributions corresponding to polymer bending, confinement, solvation and electrostatics. An approximate functional that allows for the solution of the minimum energy configuration in the strongly bound limit is shown to predict the pitch and surface area of observed helical corona phases in the literature from the surface binding energy and persistence length. This approximate functional also predicts and quantitatively describes recently observed ionic strength-mediated phase transitions of charged polymer corona at carbon nanotube surfaces. The Hamiltonian and approximate functional provide a theoretical link between polymer mechanical and chemical properties and the resulting adsorbed phase configuration, and therefore should find widespread utility in predicting corona phase structure around nanoparticles.

611256 Flame Spray Pyrolyzed Lithium Titanate ($Li_4Ti_5O_{12}$) As Promising Anode Material for Lithium-Ion Batteries

Abstract:

Spinel-structured lithium titanate ($Li_4Ti_5O_{12}$, LTO) is a one of the most attractive anode materials for next generation of lithium-ion batteries, due to its excellent stability, zero volume change during charge/discharge cycling and safe performance [1]. However, its size is crucial for optimized electrochemical performance.

In this regard, here, we synthesize LTO nanoparticles of different sizes (8-21 nm) by tuning flame spray pyrolysis (FSP) operating conditions. FSP is an attractive process for the controlled one-step synthesis of functional multicomponent oxides from low cost precursors and permits fine tuning of the nanoparticle properties by varying processing conditions [2]. Larger primary particles are synthesized by increasing precursor concentration in the flame which is attributed to longer particle residence time at high temperatures. As-synthesized LTO nanoparticles are used as active materials for the fabrication of working electrodes for lithium-ion battery half-cells. For comparison, LTO is also made by Pechini method and both FSP- and Pechini- made samples are compared with a commercial LTO powder in terms of their electrochemical properties. To evaluate their electrochemical performance, galvanostatic charge-discharge, rate performance and cycling voltammetry experiments are performed.

FSP made LTO of 18 nm shows the best electrochemical performance with almost 100% capacity retention after 500 charge/discharge cycles, whereas the smallest particle (8 nm) has a capacity loss more than 60%. Particle size < 16 nm leads to less efficient use of the intercalation sites due to higher ratio of surface to bulk atoms and to stress caused by the higher curvature at the crystallite surface. Tuning the particle size by optimization of flame spray conditions leads to formation of particles with specific capacity close to the theoretical value for LTO (175 mAh/g) which indicates maximum utilization of the active material per charging/discharging cycle.

References

1. Sun X., Radovanovic P., Cui B., Advances in spinel $Li_4Ti_5O_{12}$ anode materials for lithium-ion batteries. *New Journal of Chemistry* 2015;29:38-63.
2. Tsikourkitoudi V., Karlsson J., Merkl P., Loh E., Henriques-Normark B., Sotiriou G.A., Flame-made calcium phosphate nanoparticles with high drug loading for delivery of biologics. *Molecules* 2020;25:1747.

611401 Effect of Black TiO_2 Nanotube-Based Electrodes for Photoelectrochemical Hydrogen Generation

Abstract:

Photoelectrochemical hydrogen generation is a green and sustainable process where the hydrogen can either be used directly or to synthesize liquid fuels. In this study, black TiO_2 nanotube fabrication was optimized through electrochemical anodization and reduction in an ethylene glycol electrolyte. The surface morphology, phase crystallinity, and oxidation states were confirmed by characterization using scanning electron microscopy (SEM), X-ray diffraction (XRD) and X-ray photoelectron spectroscopy (XPS). SEM showed a uniform nanotube structure with an average pore diameter of 65nm while XRD indicated anatase crystallinity phase. XPS data demonstrated oxygen vacancies of around 17%. Light-chopped sweep voltammetry and cyclic voltammetry of the black TiO_2 nanotube suggested that the onset potential and oxidation/reduction states were -0.8V and -1.2/-0.9V respectively. The photoelectrochemical performance was investigated in a three-electrode single compartment cell using a Pt wire as a counter electrode, Ag/AgCl (saturated KCl) as a reference electrode and the black TiO_2 nanotube as photoanode. Hydrogen was detected online via residual gas analyzer (RGA) where a maximum pressure of 26.7μPa was observed over an hour. Photocurrent intensity was measured using a Gamry potentiostat and reached up to 2.26 mA/cm^2 under acidic conditions while 0.87 mA/cm^2 over alkaline conditions.

611540 Rheological Characterization of Oxidized Carbon Black Suspensions in Aqueous Salt Solutions for Use As Flowable Electrodes in Capacitive Deionization

Abstract:

Climate change, coupled with rising demand on already strained aquifers, threatens water security across the globe. Currently, 1.5 billion people lack access to clean water, and the UN projects this number to rise to 2.8 billion before the end of 2025. In coastal regions like the Middle East and California, desalination has increasingly been utilized to address water scarcity. Whereas most industrial desalination methods rely on thermal distillation or high-pressure reverse osmosis to remove salt from saline and brackish water, capacitive deionization (CDI) has the potential to reduce energy usage while retaining high salt removal, ensuring lower water costs and increased accessibility.

One class of scalable CDI technology is flow capacitive deionization (FCDI). A key component of this technology is a suspension of electrochemically active particles that are continuously fed through a device to perform capacitive deionization of seawater. FCDI promises lower energy usage and expense compared to traditional CDI methods, but the high viscosity of slurry flow electrodes currently limits practical use. To overcome this hurdle, the viscosity of these suspension electrodes must be reduced while maintaining the ability to perform electrochemical work. Carbon black is the most common conductive material within a slurry electrode, and one method to lower viscosity is oxidation of the carbon's surface.

In this work, we investigate the impact of acid-based surface oxidation on the colloidal stability of carbon black nanoparticles for use in FCDI cells. More specifically, we use dynamic light scattering and zeta potential to measure the stability ratio as a function of pH and ionic concentration and, in turn, its effect on the macroscopic rheology of carbon black in salt solutions. Under harsh oxidation conditions, we find carbon black adopts strong negative (repulsive) charge at high pH. As a result, improved colloidal stability is observed in comparison to the unoxidized form. This can be rationalized within the context of DLVO theory. Furthermore, our findings demonstrate strong surface oxidation provides a way to enhance stability in colloidal suspensions of carbon black in aqueous salt solution, thus reducing flow electrode viscosity and increasing viability for use in desalination technology.

611714 Indium-Palladium Nano-Enabled Hollow Fiber Reactor Improves Hydrogen Delivery for Nitrate Reduction to Innocuous Nitrogen Gas in Continuous Flow Operation

Abstract:

Nitrate is the most commonly occurring anthropogenic groundwater pollutant in the US, due largely to over application of nitrogen-based fertilizers. In drinking water, nitrate consumption can cause oxygen deficiency in the bloodstream; an especially dangerous occurrence for infants. Accordingly, the US EPA has imposed an MCL of 10 mg NO_3-N/L. Existing technologies to treat nitrate focus on separating nitrate from water using ion exchange or reverse osmosis that produce brine waste streams that require disposal. Also, a few full-scale treatment systems currently operate using biological denitrification for drinking water treatment wherein nitrate is converted to innocuous by-products.

Non-biological catalytic hydrogenation rapidly converts nitrate to nitrogen gas using hydrogen gas (H_2) and a solid-phase catalyst. Unlike biological systems, these abiotic catalytic hydrogenation can be turned on or off with near zero start-up time. However, 80% of publications regarding catalytic hydrogenation treatment option focus on material development and employ batch slurry reactor with a catalyst in suspension and uncontrolled hydrogen flow, which is not scalable for drinking water treatment.

The presented technology uses hollow fibers as a method of controlling hydrogen delivery in a continuous flow reactor. This method improves the efficiency of hydrogen delivered, which results in an order of magnitude decrease in required hydrogen flow rate and cost. Previously studied nitrate reduction systems have incorporated catalysts in suspension, but those configurations risk metal nanoparticles carrying over into drinking water. To resolve this technology barrier, the hollow fibers also act as a surface attachment point for the catalyst. After analyzing various catalyst coating methods, we have found a method to irreversibly attach nano-catalysts to the H_2-permeable hollow-fibers. The research focuses on a bimetallic Palladium-Indium catalyst because it has the highest selective reduction of nitrate to nitrogen gas without leaching. The parameters of the synthesized Pd-In catalyst and the reaction result in up to four times the nitrate reduction activity as similar catalysts while still achieving nitrogen selectivity above 95%. Reactor parameters and the corresponding effects on activity and selectivity of the catalyst will be discussed; particularly relating to scale up from batch to continuous flow. The reactor system is operated in both pure and natural waters for extended periods to determine any sources of long term catalyst deactivation. This presentation focuses on the development of a scalable technology for groundwater nitrate treatment with high selectivity to desired nitrogen gas and minimized catalyst deactivation over time.

599126 Morphological Influence of Thermal Treatment of Barley Straw Derived Si/C Composite for Potential Application in Energy Storage

Abstract:

Silicon (Si) is an abundant, cheap and environmentally friendly element, which is a potential replacement for graphite as the anode material in Li-ion batteries. This is due to the high theoretical capacity of silicon (3589 mAh g^{-1}) which is an order of magnitude greater than that of graphite (372 mAh g^{-1}). However, application of bare silicon as an anode material suffers from material pulverization during Li-ion cell operation. One of the proposed strategies to mitigate this problems is to coat the silicon with carbon [1,2]

The carbon material used in this study was extracted from bio-waste, namely barley straw. The barley straw was mechanically grinded to produce powder. The powder was then stirred in 37 wt% HCl for 3 hours. The mixture underwent Buchner filtration, subsequently the acid was removed from the filtrate by rotary evaporation. The remaining solid mass was oven dried over night. The treated barley powder and silicon nano particles were wet mixed in ion-exchanged water to produce a Si/C composite. The water was later removed from the mixture by heating it in a petri dish on a hotplate. The obtained solid Si/C composite was thermally treated in argon atmosphere using induction annealing over the range of 1000–2400°C.

Characterization of the thermally treated Si/C via SEM/TEM/XRD revealed several material transformations at different temperature points. The sample treated at 1000°C displayed carbon coated silicon (Si@C) with approximate 10 nm carbon coatings, as well as interstitial carbon being present between the silicon nanoparticles. At 1400°C, due to the carbon diffusion into silicon, the previously observed Si@C transformed into silicon carbide nano particles (SiC-3C). Further increase of the annealing temperature (>1800°C) promoted fusion of 3C-SiC nano particles into larger structures and simultaneously propagated the sublimation of Si atoms from the SiC surface. Sublimation of Si atoms resulted in growth of graphite coatings on the surface of SiC at temperatures of 1800-2400°C. The sample obtained at 2400°C displayed 3 polytypes of SiC: SiC-3C, SiC-6H, SiC-15R and 2 polytypes of graphite: 3R and 2H and turbostratic graphite. The sample was composed mainly of graphite flakes with embedded SiC particles (10-100 nm) which displayed ~10 nm thick graphite coating.

Preliminary electrochemical testing in Li-ion coin cells, revealed a reversible capacity of 1200 mAh g^{-1} for the Si@C sample, which also displayed an improved capacity retention after 50 cycles compared to the reference coin sell made from bare nano silicon particles.

References

[1] Y. Wen, Y. Zhu, A. Langrock, A. Manivannan, S.H. Ehrman, C. Wang, Graphene-bonded and -encapsulated Si nanoparticles for lithium ion battery anodes, Small. 9 (2013) 2810–2816. https://doi.org/10.1002/smll.201202512.

[2] M. lun Jiao, J. Qi, Z. qiang Shi, C. yang Wang, Three-dimensional Si/hard-carbon/graphene network as high-performance anode material for lithium ion batteries, J. Mater. Sci. 53 (2018) 2149–2160. https://doi.org/10.1007/s10853-017-1676-3.

595138 MOF-Derived $MgO/Mg(OH)_2$@Carbon for High Heat Release

Abstract:

Metal-organic framework (MOF)-derived metal oxide@nanoporous carbon (NC) has drawn increasing interest in recent years since nanoscale metal oxides can be well dispersed in the carbon matrix. Herein, we report the synthesis of MOF-derived $MgO/Mg(OH)_2$@NC, which shows high thermal energy release and storage through reversible chemical transformations. The reversible phase transition triggered by hydration and dehydration was confirmed by powder X-ray diffraction (PXRD) analysis. MOF-derived MgO@NC exhibits an enhanced heat release compared to commercially available MgO because of its large surface area and homogeneous distribution of nanometer-sized MgO in the NC matrix. Details of the work will be presented.

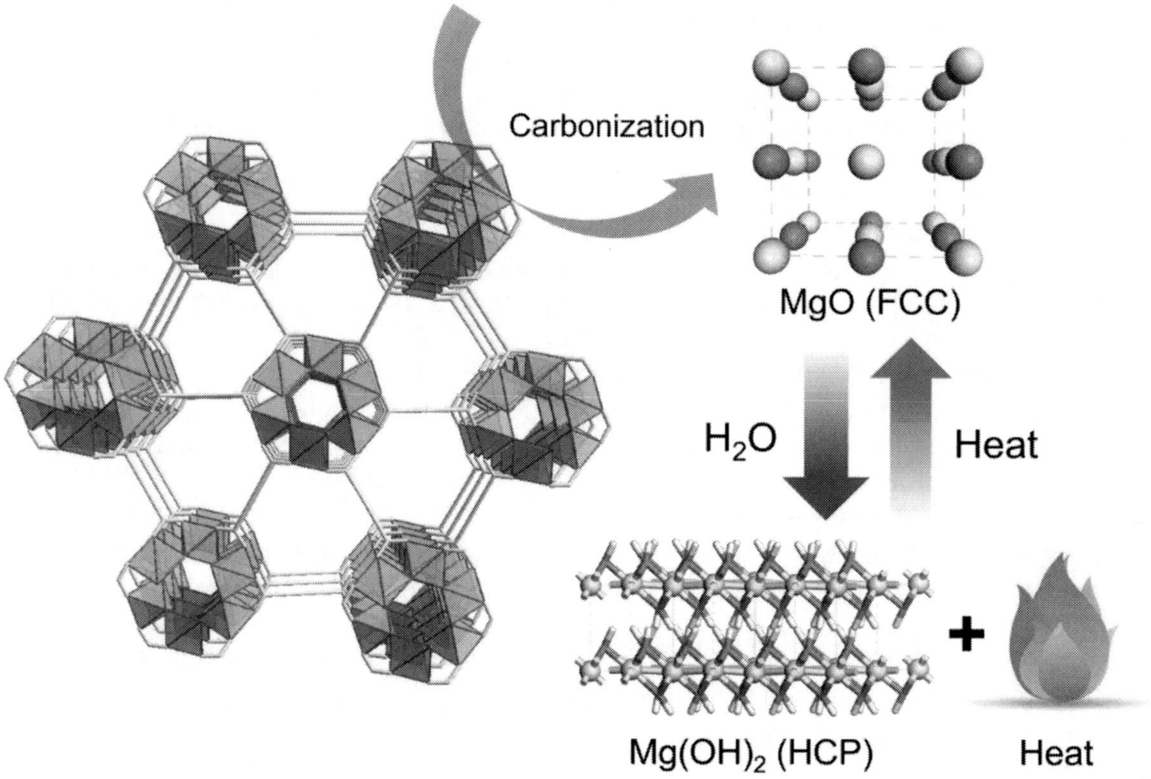

601187 Investigating the Effects of Lithium Phosphorous Oxynitride Coating on Blended Solid Polymer Electrolytes

Abstract:

Solid state electrolytes are very promising to enhance the safety of lithium ion batteries. Two classes of solid electrolytes, polymer and ceramic, can be combined to yield a hybrid electrolyte that can synergistically combine the properties of both materials. For instance, the chemical and thermal stability and high mechanical modulus of ceramic electrolytes against dendrite penetration can be combined with the flexibility and ease of processing of polymer electrolytes. By laminating a polymer electrolyte with a ceramic electrolyte, the stability of the solid electrolyte is expected to improve against lithium metal, and the ionic conductivity could remain close to the value of the original polymer electrolyte as long as an appropriate thickness of the ceramic electrolyte is applied. Here we report a bilayered lithium-ion conducting hybrid solid electrolyte consisting of a blended polymer electrolyte (BPE) laminated with a thin layer of the inorganic solid electrolyte lithium phosphorous oxynitride (LiPON). The hybrid system is thoroughly studied. First, we investigated the influence of polymer chain length and lithium salt ratio on the ionic conductivity of the BPE based on poly(ethylene oxide) (PEO) and poly(propylene carbonate) (PPC) with the salt lithium bis(trifluoromethanesulfonyl)imide (LiTFSI). The optimized BPE consists of 100k molecular weight PEO, 50k molecular weight PPC, and 25(w/w) % LiTFSI, (denoted as PEO100PPC50LiTFSI25) which exhibits an ionic conductivity of 2.11×10^{-5} S/cm, and the ionic conductivity shows no thermal memory effects as the PEO crystallites are well disrupted by PPC. Secondly, the effects of LiPON coating on the BPE were evaluated as a function of thickness down to 20 nm. The resulting bilayer structure shows an increase in the voltage window from 5.2 to 5.5V (vs Li/Li^{+}) and thermal activation energies that approach the activation energy of the BPE when thinner LiPON layers are used, resulting in similar ionic conductivities for 30nm LiPON coatings on PEO100PPC50LiTFSI25. Coating BPEs with a thin layer of LiPON is shown to be an effective strategy to improve the long-term stability against lithium of the developed solid-state lithium ion conductors.

602045 Layered Double Hydroxide As Co-Catalyst to Improve the Oer Kinetics of Hematite Photoelectrode

Abstract:

Semiconductor based photoelectrodes to produce hydrogen from water in the presence of sunlight is considered as one of the most promising solution to overcome the energy crisis and global warming the world is facing today. The basic requirements for a semiconductor to be used in photoelectrochemical (PEC) water splitting are efficient absorption of visible light, high stability, low overpotential and good charge transport. Over the years, hematite (α-Fe_2O_3) based photoelectrodes have received a lot of attention because of its non-toxicity, availability, low cost, high photochemical stability in basic media and narrow bandgap (1.9 to 2.2 eV). From the perspective of scalability, hematite (α-Fe_2O_3) holds immense potential because of its high theoretical solar to hydrogen efficiency (~ 15 %). However, its application in PEC water splitting suffers from several limitations such as comparatively low absorptivity, very short excited-state lifetime (10^{-12} s) and short hole diffusion length (2 to 4 nm). The resulting high recombination of photogenerated electron-hole pairs and slow oxygen evolution reaction (OER) kinetics need to be improved. Herein, we report the synthesis of Sn doped hematite and its integration with NiFe layered double hydroxide (LDH) as co-catalyst to overcome the aforementioned drawbacks. $SnCl_2$ at varying concentrations is used as the Sn source to prepare Sn doped hematite nanostructures. Furthermore, NiFe LDH is coupled to Sn doped hematite using Ni and Fe precursors. The nanostructured photoelectrodes were investigated by XRD, SEM with EDX, UV-VIS and the Sn elemental doping was confirmed by XPS analysis. The cocatalyst loaded sample exhibited a much higher photocurrent density ca. 2.2 $mAcm^{-2}$ at 1.75 V vs. RHE compared to 0.2 $mAcm^{-2}$ for Sn doped hematite at the same potential. Electrochemical Impedance Spectroscopy analysis further confirms that addition of cocatalyst overlayer helps in enhancing the surface reaction kinetics and reducing the OER barrier.

598527 Colorimetric Response of Borohydride Stabilized Silver Nanoparticle on Interaction with Organophosphates

Abstract:

Detection of residual organophosphates in food is of paramount importance in the current scenario. Biosensors for pesticide detection may be developed by monitoring the interaction of nanoparticles with pesticides. In this report, we have studied the colorimetric response of pesticides phorate, chlorpyrifos, and malathion with borohydride stabilized silver nanoparticle. The colorimetric changes are corroborated with UV-Visible spectra along with the change in particle size and zeta potential. The decolorization of phorate with nanoparticles was explained. The reaction mechanism was proposed by the formation of silver and sulfur atom (Ag→S) complex, which is verified by Raman spectroscopy data of the systems. The uncapped nanoparticles get stabilized by pesticides over time for chlorpyrifos and malathion while phorate undergoes degradation due to scission of S-CH_2 present in the molecule. The color change may be utilized as a colorimetric method for the detection of phorate pesticide.

Colorimetric Response of Bare Silver Nanoparticle with pesticide

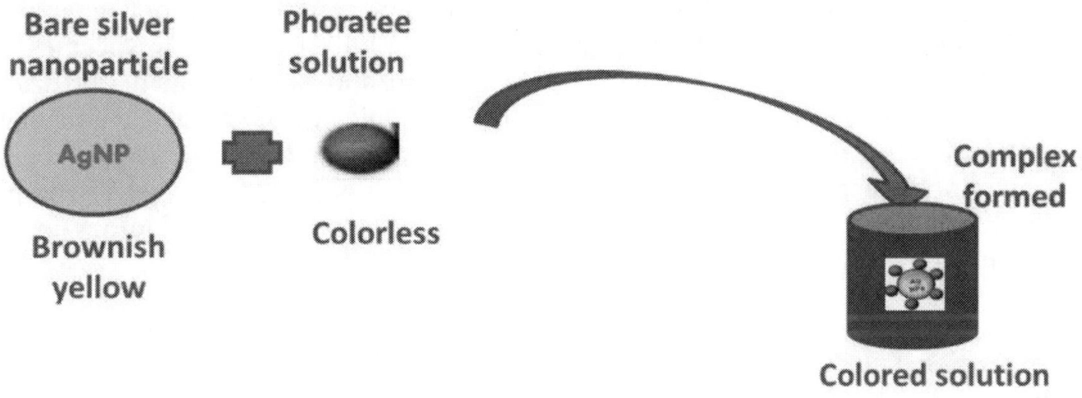

602182 Monolithic Chromatographic Purification of Plant Viral Nanoparticles

Abstract:

Commercial biomanufacturing is being challenged by the emergence of larger biomolecule products. Biomedicals sectors such as gene therapy, vaccine development, and medical imaging have research pipelines rich with biological product platforms based on viral nanoparticles (VNPs) and other larger biomolecules. Traditional downstream unit operations, such as resin-based liquid chromatography columns, are often still used for purification of these large biomolecules. In these situations, process performance is significantly impaired when compared to the intended use case of protein separation. The low diffusivity of larger biomolecules within the resin results in slow chromatographic binding and elution kinetics, which in turn results in reduced binding capacity, resolution of separation, and process throughput.

Monolithic chromatography represents an alternative solution to larger biomolecule purification. Mass transport is dictated by convective flow in monolithic chromatography, whereas transport in resin-based chromatography is largely controlled by diffusion in the resin bead pores where the functional ligands reside. Monolithic chromatography has been successfully employed for purification of larger biomolecules including VNPs based on adeno-associated viruses, influenza viruses, lentiviruses, and bacteriophages. However, purification of plant VNPs has not been previously established in literature. Plant VNPs are currently being explored for utility in vaccines, medical imaging contrast agents, drug delivery vehicles, biosensors, and pharmaceutical purification. They will continue to play a key role in the advent of larger biomolecule products – upstream production is simple, inexpensive, and scalable using plant-based cultivation and they are non-pathogenic in humans, an important safety consideration in commercialization. As of now, purification of plant VNPs is primarily reported using resin-based chromatography or molecular biology-based unit operations including sucrose gradient-based ultracentrifugation.

We are investigating the use of monolithic chromatography in the purification scheme of *Tobamovirus*-based plant VNPs to improve and de-bottleneck current manufacturing strategies, which are limited in throughput by low flowrates and recovery of resin-based chromatography or the low capacity and long operational time of ultracentrifugation. We will describe our process development of this novel strategy for purification of plant VNPs.

610700 A Simple Evaporative Deposition-Photopolymerization Approach for Facile Fabrication of Polymeric Hydrogel Films Containing Micropatterned Opal Structures

Abstract:

A Simple Evaporative Deposition-Photopolymerization Approach for Facile Fabrication of Polymeric Hydrogel Films Containing Micropatterned Opal Structures

Subhash Kalidindi, Maurice Bukenya, and Hyunmin Yi

Department of Chemical and Biological Engineering, Tufts University, Medford, MA-02155, USA

Abstract:

Artificial opal structures have gained significant attention towards optical applications for sensing and detection due to the ready tunability and control of colors in response to stimuli.[1] In specific, opal structure-based hydrogel platforms have shown great potential due to the reversible volume change they offer in response to external stimuli (e.g. temperature, pH, ionic strength).[2-5] However, existing techniques to fabricate these hydrogel platforms suffer from long opal deposition times and involve complicated procedures.[6]

In order to address these challenges, we exploit a simple, reliable and robust fabrication technique: micromolding-based evaporation-polymerization method for fabrication of poly (ethylene glycol) (PEG)- based hydrogel films having micropatterned opal structures. First, deposition of uniform opal structures with intense colors is achieved by evaporation of polystyrene (PS) nanospheres in patterned polydimethylsiloxane (PDMS) micromolds. The uniformly deposited opal structures exhibit intense color via darkfield optical microscopy, and the wavelength of the opal color is confirmed using visible reflectance spectroscopy. Further, scanning electron microscopy (SEM) confirms uniform close hexagonal packing. The opal structures are then captured into hydrogel films via simple photopolymerization of PEG diacrylate monomers using a hand-held UV lamp. Dark field microscopy and SEM confirm reliable and consistent capture of the opal structure and robust nature of the films. By tuning simple parameters like monomer concentration and PS nanosphere size, wide color range in the visible spectrum is readily achieved, confirming rapid, reliable and robust nature of our simple evaporation-polymerization technique. Finally, incorporation of functional monomers like acrylic acid (AA) and methacrylic acid (MAA) leads to stimuli responsive opal-hydrogel films showing potential applications in the detection of pH and ionic strength. We envision that the simple evaporation – polymerization technique can be readily extended to manufacture a variety of functional materials for facile monitoring of environmental conditions.

References:

1. Armstrong, E. and C. O'Dwyer, Artificial opal photonic crystals and inverse opal structures– fundamentals and applications from optics to energy storage. Journal of materials chemistry C, 2015. 3(24): p. 6109-6143.
2. Alexeev, V.L., et al., High ionic strength glucose-sensing photonic crystal. Analytical chemistry, 2003. 75(10): p. 2316-2323.

3. Lee, K. and S.A. Asher, Photonic crystal chemical sensors: pH and ionic strength. Journal of the American Chemical Society, 2000. 122(39): p. 9534-9537.
4. Shin, J., P.V. Braun, and W. Lee, Fast response photonic crystal pH sensor based on templated photo-polymerized hydrogel inverse opal. Sensors and Actuators B: Chemical, 2010. 150(1): p. 183-190.
5. Wang, J., et al., Multiresponsive hydrogel photonic crystal microparticles with inverse-opal structure. Langmuir, 2013. 29(28): p. 8825-8834.
6. Yu, B., et al., Preparation of Humidity-Sensitive Poly (Ethylene Glycol) Inverse Opal Micropatterns Using Colloidal Lithography. Materials, 2017. 10(9): p. 1035.

598416 (Invited Plenary Talk) Wearable Electrochemical Sensors

Abstract:

Wearable sensors have received a major recent attention owing to their considerable promise for monitoring the wearer's health and wellness. The medical interest for wearable systems arises from the need for monitoring patients over long periods of time. These devices have the potential to continuously collect vital health information from a person's body and provide this information to them or their healthcare provider in a timely fashion. Such sensing platforms provide new avenues to continuously and non invasively monitor individuals and can thus tender crucial real-time information regarding a wearer's health. Our research over the past decade has aimed toward filling the gaps toward providing biochemical information, beyond that given by common wrist-watch mobility trackers. This presentation will discuss recent developments in the field of wearable electrochemical sensors integrated directly on the epidermis or within the mouth for various non-invasive biomedical monitoring applications. Particular attention will be given to non-invasive and minimally-invasive monitoring of metabolites and electrolytes using skin-worn and microneedle amperometric and potentiometric sensors, respectively, along with related materials, energy and integration considerations. The preparation and characterization of such wearable electrochemical sensors will be described, along with their current status and future prospects and challenges.

610185 (Invited Plenary Talk) an Integrated Paradigm for Precision Exposure to Airborne Chemical and Biological Stressors Based on Personal Sensing

Abstract:

Introduction

The current manuscript outlines the fundamentals of a methodological framework for precision exposure assessment and gives concrete examples of its application in different European cities to assess the health risk associated with exposure to fine and ultra fine particles of chemical and biological origin. The work described herein involves a multi-city personal sensor campaign aiming to refine PM exposure using low-cost portable sensor data, exposure and human respiratory tract deposition modelling to estimate the biologically effective internal daily intake on an individual basis and agent-based modelling to estimate exposure and consequent health risk to the community.

Methodology

A prototype monitoring device (IcaruSense) was developed for measuring three size fractions of PM (1, 2.5 and 10 μm), enabling direct assessment of personal exposure. The device is based on an Arduino microcontroller where miniaturized electrochemical sensor-modules are connected. In order to improve sensor performance, correction factors were used to account for diverse environmental conditions and they were calibrated with the help of a well-validated light scattering optical sensor. Study participants wore a physical activity wristband (Garmin Vivosmart 3) that records steps, distance, type of activity, heartbeat and sleeping patterns. Finally, participant positions were recorded using a GPS sensor, integrated into the PM sensor kit. After calibration and validation, the IcaruSense devices were used to capture daily variability of PM exposure. Exposure was further refined by estimating inhalation adjusted intake rate from the continuous time-activity data collected from the physical activity and the GPS sensors. Using the inhalation adjusted intake rate PM deposition across the human respiratory tract (HRT) was readily estimated using the Multiple Path Particle Deposition (MPPD) model.

A city scale agent-based model (ABM) was developed to support this work by expanding the precision exposure estimates from the individual to the community level. The model feeds into population-based exposure assessment without imposing prior bias, basing its estimations onto emergent properties of the behaviour of the computerised autonomous decision makers (agents) that compose the city-system. Population statistics, road and buildings network data were transformed into human, road and building agents, respectively. Informed by literature-extracted relationships between Socio-Economic Status (SES) and human behavioural patterns, the established ABM simulates societal dynamics. Time-use survey outputs were associated with human agent behavioural rules, aiming to model representative to real-world routines. Moreover, real time-geography of exposure data, extracted from the precision sensors campaigns in nine European cities were used to parametrise and enhance the model.

Mobility behaviour and eventually exposure to pollutants change as a result of a human agent-specific decision making. Virtual individuals of different sociodemographic backgrounds use different means of transportation and follow different sequence and types of activities. Behaviours that were not explicitly programmed, arise through the human agents' interactions, enabling the examination of emergent behaviours from the bottom-up. Spatiotemporal trajectories are coupled with spatially resolved Particulate Matter (PM) levels, enabling the assessment of personal and inhalation adjusted exposure, based on type of location and intensity of enrolled activities.

Results and Discussion

The above methodology was applied in the personal sensors campaign of the HORIZON2020 EU Project ICARUS, where exposure and intake to PM of almost 100 individuals in each out of nine European cities were monitored. The integrated hybrid approach outlined above, which couples multi-sensor datasets on an individual basis with advanced respiratory tract modeling, allowed us to significantly differentiate the actual intake of the participants, highlighting larger differences than the ones attributed to the spatial differentiation of air pollution based on fixed stations (difference of ambient PM levels of 50% were translated in intake differences up to 110%). These differences are the result of the differences in PM size fractions that are captured by the sensors and the capability of the HRT model to translate these differences in PM size distribution accounting for variation in respiratory tract physiology among the participants, which is ultimately reflected in intake estimates.

ABM findings indicated that inhalation adjusted exposure differences between 2 individuals living in the same neighbourhood can vary by as much as 87%, due to radically different spatiotemporal behaviours. This major difference was observed for cases where one human agent has a full-time job and often exercises in outdoor courts whereas his/her neighbour is a retired homemaker. For the same reason, ABM runs confirmed that even people who reside in the same dwelling might not necessarily be exposed to similar levels of pollution. Model results indicated that PM inhalation adjusted exposure between flatmates can differ by 61%. Evidence was, also, provided for vulnerable subgroups of population that are disproportionately subjected to high levels of exposure. ABM-retrieved population exposure findings indicated that subgroup median exposure for children, adult males with a full-time job and low income and human agents of a lower educational level was notably higher than the total population median value. The validated model can assess exposure for the entire sociodemographic spectrum of a simulated region, without the constant need for repeated individual monitoring which is often not feasible due to high costs, ethical constraints and complex organisation that is associated with such measurements. By modelling the heterogeneous routines of human agents, the ABM produces detailed information related to the societal system examined and generates data that could be used to fill in the gaps that exist in traditional datasets.

Conclusions

This study presents a new methodological approach to estimate the external exposome, which encompasses the totality of human environmental exposures at an individual and community level. This method helps to integrate substantive considerations of individual status into long-term urban planning and management; it would also facilitate the identification of both physically and socially practical means for reducing life-threatening exposure levels. The methodological paradigm for precision assessment of exposure at the individual and community levels laid out in this work can be used for evaluating the impacts of different public health policies prior to implementation.

611204 Synergistic Effect of High Sulfur Loading Layered Cathode, Ceramic Separator and Gel Electrolyte

Abstract:

Lithium sulfur (Li-S) batteries are promising candidates for next-generation energy storage devices as they are capable of providing higher energy density and lower cost in comparison to the state of the art Li-ion batteries. To gain these characteristics, there are challenges such as polysulfide crossover, large volume changes during cycling, and deposition of insulating lithium sulfide that need to be alleviated. A popular approach to address these issues is to control the structure and properties of the carbon cathode in order to influence the transport of polysulfides, surface area for reaction, and lithium sulfide deposition. Ceramic separators on the other hand can drastically affect ion diffusion and overall battery performance. Safety concerns regarding the flammable liquid electrolyte can be reduced by deploying gel electrolyte.

Towards this direction, we developed a layer-on-layer cathode structure with sulfur-impregnated activated carbon alternating with graphene layer, which was fabricated by Air-assisted electrospray. A much higher sulfur loading compared to slurry cast cathodes can be provided in this method. Combining the new cathode with gel electrolyte can offer higher safety of the battery and capacity retention. Ceramic separator would compensate the reduced ion conductivity of the gel electrolyte and the combination of these variables will be very promising in improvement of battery performance and safety.

608371 TiO_2 Nanoparticles Wrapped with Short Single Wall Carbon Nanotubes for Enhanced Photocatalytic Performance

Abstract:

The interfacial contact between TiO_2 and carbon-related structures is crucial in monitoring electron transfer rate and, in turn, the photocatalytic efficiency. Herein, we report a facile approach to wrap short single wall carbon nanotubes (SWCNTs) on TiO_2, with improved heterojunctions and delayed charge carrier recombination. Ultrasonication-assisted cutting was utilized to cut long SWCNTs (1 – 3μm in length) to short SWCNTs (50-400 nm in length). The short SWCNTs were wrapped on TiO_2 nanoparticles (100 nm) via a hydration-condensation technique. The synthesized nanostructures were characterized by X-ray diffraction (XRD), Raman spectroscopy, scanning electron microscopy (SEM), transmission electron microscopy (TEM) and photoluminescence spectroscopy. In comparison to traditional TiO_2-SWCNT composites, TiO_2 wrapped by short-SWCNTs showed longer lifetimes of photogenerated electrons and holes, as well as a higher photocatalytic activity in gas-phase degradation of acetaldehyde. In addition, upon comparison with a TiO_2-nanographene "quasi-core-shell" structure, TiO_2-SWCNT structures offer better electron capturing efficiency and higher photocatalytic performance, revealing the impact of the dimensions of graphitic structures on the interfacial transfer of electrons and light penetration to TiO_2. The engineering of TiO_2-SWCNT structure is expected to benefit photocatalytic degradation of other volatile organic compounds, and may provide alternative pathways to further improve the efficiency of other carbon-based photocatalysts.

610165 A Simulation Study of the Effect of CO on the Iron Particle and Nanotube Nucleation.

Abstract:

Single-walled carbon nanotubes (SWCNTs) are seamless cylinders of graphene that have been at the forefront of nanotechnology research for the past two decades. They possess a range of exceptional properties including high strength (~37 GPa), thermal conductivity (~3500 W/m/K), and ballistic electronic transport. Recent progress in FC-CVD using Iron particles has shown the ability of these particles to produce highly selective chiral nanotubes with diameter profiles in the range of 1 nm to 4 nm. Iron (Fe) nanoparticles have attracted great interest due to their potent magnetic and catalytic properties which strongly depend on the structures and morphologies.

In this work, we use molecular dynamics to simulate a floating catalyst chemical vapor deposition (FC-CVD) process using carbon monoxide (CO) as the main carbon source. Additionally, DFT simulations help to quantify the strength of interaction and study preferred adsorption sites for different surface compositions. Detailed knowledge of the structure and dynamic evolution in the composition for the iron particles during the production of carbon nanotubes (CNTs) is important for understanding many surface phenomena such as adsorption, oxidation, and catalytic reactions and their effects on nanotube nucleation and growth.

599594 Molecular-Dynamics Analysis of the Thermomechanical Behavior of Nanodiamond Superstructures in Interlayer-Bonded Twisted Bilayer Graphene

Abstract:

Graphene derivatives and metamaterials are usually fabricated through chemical functionalization and defect engineering of graphene sheets. These materials have potential for numerous technological applications since tailoring their nanostructure enables fine tuning of their thermomechanical properties. Toward this end, we have conducted a systematic computational study on the thermomechanical behavior of a class of two-dimensional (2D) graphene-diamond nanocomposite superstructures, formed through patterned hydrogenation-induced interlayer covalent bonding of twisted bilayer graphene with commensurate bilayers. These superstructures are fully characterized by the commensurate bilayer's twist angle, the interlayer bond pattern, and the concentration of sp^3-bonded C atoms in the nanocomposite material. The analysis of their thermomechanical behavior is based on molecular-dynamics (MD) simulations according to a reliable interatomic bond-order potential.

We report results on the mechanical behavior of such carbon nanocomposite superstructures based on MD simulations of uniaxial straining tests and establish the dependence of the superstructures' mechanical properties on the concentration of sp^3-bonded C atoms. We also demonstrate that a brittle-to-ductile transition occurs in these superstructures with increasing the concentration of sp^3-bonded C atoms beyond a critical level. The underlying ductile fracture mechanism, mediated by void formation, growth, and coalescence, is characterized, and the superior mechanical response to uniaxial straining of the ductile nanodiamond superstructures is demonstrated.

Furthermore, we study the mechanical response of these superstructures to indentation and shear straining based on MD simulations of nanoindentation tests and dynamical shear straining tests. We establish the dependences of the elastic modulus, hardness, and shear strength of the superstructures on the full range of their structural parameters, especially on the fraction of diamond in the nanocomposite superstructure measured by the concentration of sp^3-bonded C atoms. The resulting structural responses and fracture mechanisms of the superstructures under nanoindentation testing and shear straining also are characterized in detail. Finally, we report results for the lattice thermal conductivity of these superstructures based on non-equilibrium MD simulations of thermal transport. We find that the lattice thermal conductivity is reduced significantly with increasing concentration of sp^3-bonded C atoms in the superstructures, which makes these 2D carbon nanomaterials very promising for thermal management applications.

600982 Nucleation and Growth of Solid-Electrolyte Interphase Components at Electrode Surfaces: Role of Electrolyte Solution Chemistry and Concentration

Abstract:

Solid Electrolyte Interphases (SEI) grow at the electrode surface because of electrolyte chemical and electrochemical instability. The SEI nucleation and growth start even at open circuit conditions, especially with anode surfaces as reactive as Li, Na, and other metals, when they are in contact with liquid or solid electrolytes. At open circuit conditions, the reaction can be described as Li oxidation and electrolyte reduction. In this talk I will discuss this spontaneous SEI formation and especially the role of the electrolyte chemistry and concentration on the initial nucleation and posterior growth. Our methods include density functional theory, classical molecular dynamics, ab initio molecular dynamics, and constrained molecular dynamics simulations. The computational results are discussed in the context of experimental observations.

597901 Inorganic-Organic Composite Electrodes for Supercapacitor Application

Abstract:

Nickel cobaltite composite electrodes were fabricated by a fast, scalable, and cost-effective electrophoretic deposition for supercapacitor applications. NiCo2O4/PANI/rGO composite electrode went through heat treatment under nitrogen to carbonize PANI after electrophoretic deposition. NiCo2O4 platelets, distributed on carbonized polyaniline–rGO network, were active in charge storage with high capacitance, excellent rate capability (1235 F g−1 at 60 A g−1), and acceptable cycling stability (3000 cycles at 10 A g−1) in a three-electrode assembly. The practical application of the composite electrode was investigated by an all-solid-state asymmetric supercapacitor cell using NiCo2O4/C-PANI/rGO as the cathode and activated carbon as the anode with specific capacitance of 262.5 F g−1 at 1 A g−1 and a good capacitance retention of 78% after 3500 cycles at an expanded working potential of 1.5 V. This work shows the importance of the composite assembly process that governs the microstructure of the composite. In addition, the effect of oxide pore tuning on charge storage is investigated.

605989 Overcoming Intrinsic Energy Density Deficiencies in Supercapacitors By Embedded Electrode Doping

Abstract:

Developing new materials to enhance the energy storage capabilities of supercapacitors is pertinent to the advancement of the energy industry. Supercapacitors have been successfully developed to increase the energy density of capacitors by adding Faradaic components to the electrodes, either by the inclusion of conducting polymers or metal oxides. Faradaic energy storage is superior in energy density to charge separation based energy storage, although by incorporating both mechanisms, supercapacitors can store energy that can be released at high powers with improved energy densities. Cycle life for surface-bonded Faradaic components is a prominent hurdle, where the energy density advantages are only sustainable for ~50 – 100 cycles. Therefore in this work we incorporate our Faradaic components within the electrodes, trapping them sterically within. This permits the augmented energy density to persist for 1000 cycles and due to the porosity of the electrodes the specific capacitance is still > 300 F/g in multi-walled carbon nanotube based electrodes. These electrodes have additional benefits in enhanced heterogeneous catalysis, which has been utilized to increase the power density of redox flow batteries as well. Further development of the energy density of supercapacitors in important for future energy storage markets as the world enters a new phase of energy.

608322 Unique Mixing Behaviors of Nanoscale Fluid-Based Electrolytes for Sustainable Energy Storage

Abstract:

Although the costs associated with renewable energy technologies continue to decrease at a fast rate, they continue to face intermittency and storage challenges. It is now possible to store this excess renewable energy in the form of chemicals and fuels through electrochemical CO2 reduction or in the form of redox flow batteries (RFBs). The electrochemical conversion of CO2 remains challenged by the CO2 solubility in aqueous solutions and the energy density of redox flow batteries is limited by redox species solubility. Therefore, novel reactive materials are currently being explored as additives to aqueous and non-aqueous electrolytes in order to improve target species solubility and/or enhance reaction rates. Polymer functionalized nanoparticles are a promising class of electrolyte additives due to their thermal stability and chemical tunability. Interestingly, grafted polymers have been shown to change their conformations in response to their environment (temperature, pH, salt concentration), thus allowing them to be engineered to capture and subsequently release target species. In this study, Nanoparticle Organic Hybrid Materials (NOHMs)-based fluids have been designed as hybrid electrolyte systems for energy storage applications. The development of NOHMs-based electrolytes requires a fundamental investigation into the structuring and organization of these complex nanoscale fluids. A study of the mixing behaviors of NOHMs with a variety of secondary fluids has revealed that physical properties such as hydrodynamic size, viscosity and diffusion coefficient are highly dependent on the chemical nature and size of the secondary fluid molecule. Additionally, it was discovered that the viscosity and the diffusion coefficient of the untethered polymer were found to scale according to theoretical predictions for the semi-dilute unentangled regime, while this was not observed for NOHMs-based fluids. Lastly, the CO2 solubility of NOHMs-based electrolytes has been measured and a change in the selectivity of the CO2 reduction reaction has been identified in the presence of these nanoscale electrolytes. The results of this work provide important insights into the design and development of a novel reactive medium for energy storage applications.

600178 Facile and Ultrafast Microwave Synthesis of $NiCo_2S_4$/Graphene Nanocomposite for High Performance Supercapacitor

Abstract:

Binary metal sulfides, especially nickel-cobalt sulfides ($NiCo_2S_4$), and their composites have attracted considerable attention as electrode materials for supercapacitors because of their excellent electronic conductivity and high capacitance compared to mono-metal sulfides and oxides. A two-step hydrothermal process is widely used to synthesize $NiCo_2S_4$. However, the traditional hydrothermal method has several drawbacks such as low energy efficiency, long reaction time and complex synthesis steps. Herein, we present an ultrafast microwave approach for fabricating $NiCo_2S_4$/graphene nanocomposite used for supercapacitor electrodes with excellent electrochemical performance. The morphologies, crystalline structures and elementary composition of the obtained nanocomposites were characterized by SEM, XRD, FTIR, and EDS. It is found that $NiCo_2S_4$ nanoparticles can uniformly grow on the surface of graphene after tens of seconds microwave irradiation. The presence of graphene not merely enhances the conductivity of nanocomposites but also provide the transport paths for ions in electrolytes and help to reduce energy loss, thus leading to enhanced electrochemical performance. Electrochemical results indicate that the specific capacitance of as-prepared $NiCo_2S_4$/graphene nanocomposite can reach up to 656 F g^{-1} at 0.5 A g^{-1}. In addition, the $NiCo_2S_4$/graphene nanocomposite exhibits excellent cycling stability with capacitance retention over 80% even after 10000 cycles at 4 A g^{-1}. Thus, the current work provides an ultrafast and energy-efficient microwave method to synthesize high-performance $NiCo_2S_4$/graphene nanocomposite for supercapacitor applications.

610567 Investigating Ion Intercalation Mechanisms in Mxene Structures through Molecular Dynamics Simulations

Abstract:

Supercapacitors are electrical energy storage devices that have garnered much interest in recent decades due to their high power density in comparison to conventional batteries[1]. Though they play important roles in various fields, the lower energy density of supercapacitors limit their widespread application. In an effort to enhance the energy density of supercapacitors, various novel materials have been synthesized and studied as electrodes. MXenes, a new family of 2D transition metal carbides, have attracted particular interest due to their excellent pseudocapacitance originating from redox reactions at the interface.[2] Studies have shown that the process of ion intercalation can significantly enhance the specific capacitance of MXenes[3]. As a result, it is of fundamental importance to understand the microscopic behavior of ions under the confinement of MXenes.

Through a series of molecular dynamics simulations, we study the atomistic interactions between ions and the MXene surface throughout the process of ion intercalation. By analyzing various structural and transport properties, such as the coordination number and the dynamics of the solvation cages, we aim to understand how ion intercalation affects the interlayer spacings of MXenes. Additionally, the constant potential method is utilized to study how ion behavior is impacted by potentials of varying magnitude. Overall, a detailed understanding of ion solvation behavior at a wide range of potentials can further guide the design of MXene supercapacitors that display simultaneous high power density and energy density.

[1] Miller J R, Outlaw R A, Holloway B C. Graphene Double-Layer Capacitor with Ac Line-Filtering Performance. Science, 2010, 329(5999):1637-1639

[2] Lukatskaya M R, Mashtalir O, Ren C E, et al. Cation Intercalation and High Volumetric Capacitance of Two-Dimensional Titanium Carbide. Science, 2013, 341(6153):1502-1505

[3] Li J, Yuan X, Lin C, et al. Achieving High Pseudocapacitance of 2d Titanium Carbide (Mxene) by Cation Intercalation and Surface Modification. Advanced Energy Materials, 2017, 7(15):1602725

600181 Highly Flexible, Strong and Conductive Cellulose Nanofibrils/PEDOT:PSS Nanopapers for All-Solid-State Supercapacitors

Abstract:

Recently, cellulose nanofibrils (CNF) produced from cellulosic materials (e.g. wood, agricultural and industrial waste, etc.) have received increasing attention due to their unique properties such as high specific surface area, high elastic modulus, high thermal stability, as well as renewability and biodegradability. These unique properties make CNF as promising building blocks for the manufacture of many sustainable nanomaterials.

In this work, we firstly prepared CNF from pulp mill sludge by a sustainable process of formic acid (FA) hydrolysis pretreatment and the followed microfluidization. It was found that the mild FA hydrolysis (at 95 ºC for 3-6 h) pretreatment could hydrolyze most of hemicellulose, swell and break down the cellulose fibers, and the collected cellulosic solid residue with a high yield (over 75%) could be further converted to CNF with relatively low-intensity microfluidization (only two passes). Then, the CNF was coated with PEDOT:PSS *via in situ* polymerization. Afterwards, flexible and conductive CNF/PEDOT:PSS nanopapers were prepared from the above obtained PEDOT:PSS-coated CNF suspension by a simple vacuum filtration approach. Furthermore, CNF/PEDOT:PSS nanopapers were treated with dimethyl sulfoxide (DMSO) to increase the conductivity. Results showed that the DMSO-treated CNF/PEDOT:PSS nanopapers had high tensile strength (over 60 MPa), high thermal stability, and low sheet resistance (18 Ω sq^{-1}). Finally, a symmetric all-solid-state supercapacitor was assembled using the DMSO-treated CNF/PEDOT:PSS nanopapers as electrodes. It was found that the assembled supercapacitor could deliver the maximum areal specific capacitance of 888.7 mF cm^{-2} (corresponding to 111.1 F cm^{-3}) and offer the highest areal energy density of 79.0 μWh cm^{-2} (corresponding to 9.9 mWh cm^{-3}), which are among the highest values reported for PEDOT:PSS based supercapacitors. More importantly, the assembled supercapacitor showed remarkable cycling stability with the capacitance retention of 109.5% after 10,000 charge/discharge cycles at a current density of 20 mA cm^{-2}. Considering the sustainable and facile preparation process and excellent electrochemical performance, the obtained CNF/PEDOT:PSS nanopapers could be a promising candidate for flexible energy storage devices.

608197 Electrospun Infrared-Transparent Solar-Reflecting Nanofibers for Direct Radiative Cooling

Abstract:

Radiative cooling is a promising way to alleviate urban heat island effects and decrease the energy required for occupant thermal comfort. Among many emerging approaches, the use of infrared-transparent films and fabrics that allow objects to directly radiate heat through bands of atmospheric transparency while blocking solar irradiation are particularly well suited for seasonal and other temporary uses. These covers have been widely based on polyethylene due to its simple chemistry and consequently intrinsically low IR absorption. Here, we investigate the effects of polymer nanofiber morphology on the scattering and transmission properties of polyacrylonitrile (PAN) based films. By controlling the electrospinning process, a beaded fiber morphology can be fabricated which features ellipsoidal beads connected via thin cylindrical fibers. These beaded fibers exhibit greater solar scattering than either spherical or cylindrical polymer structures, as confirmed using electromagnetic simulations. These noteworthy scattering properties decrease the amount of material needed to reach above 95% solar-weighted total reflectance, enabling high infrared transmittance (> 70%) despite PAN's intrinsic absorption in the 8-13 μm range. When scaled up and tested outdoors, the beaded nanofiber PAN film (nanoPAN) decreases the temperature of an unpolished PDMS-coated aluminum sheet by 7°C by shielding it solar radiation while allowing it to radiate heat to space. Heat transfer modeling agrees with the outdoor experiments and predicts temperature drops exceeding 10°C below ambient under standard sky conditions with an optimized beaded nanofiber film. This work on tailoring the morphology of polymer nanofibers paves the way for unconventional materials and lower-purity feedstocks to be used as solar filtering covers in direct radiative cooling applications.

612443 Understanding the Role of Small Molecules in Mineral Oxide Processing By Atomic Layer Deposited Titanium Dioxide Surfaces and Colloidal Probe Atomic Force Microscopy

Abstract:

Small molecule polyelectrolytes such as polyphosphates, citric acid and Dispex N40 are widely used in mineral processing. These polyelectrolyte salts are reasoned to provide electrostatic stabilisation (possessing high surface charge densities); however this is mediated by the solution conductivity (Debye screening length). Moreover, high charge densities ($\zeta \sim 50$ mV) on mineral oxide surfaces can be achieved through pH moderation, therefore the question remains as to why small molecule polyelectrolytes are so popular within mineral processing operations?

Atomic layer deposition (ALD) was used to fabricate flat and smooth (RMS 0.5 nm) TiO_2 surfaces. With surface charges varying from 0 to -45 mV (corresponding to pH 4.7 to pH 10.5 respectively), the interaction force between TiO_2-TiO_2 (colloid probe and surface) showed jump-in to contact at all pH values. When coated with SHMP, the interaction forces were purely repulsive, with no jump-in on approach or pull-off during retraction. Such behavior was consistent over a wide pH range which included the $pH_{i.e.p.}$ for SHMP (pH 3.4) and highly charged surfaces (pH 11).

The approach data was modelled by both off-set electrical double layer forces (DLVO) and steric forces (Alexander-de Gennes model). The steric interaction was found to depend on pH with a step-change behavior in the region of pH 7.5. Below pH 7.5 the size of the SHMP molecule was ~1.0 nm and above pH 7.5 the size slightly increased to ~1.5 nm, see Fig. 1. Molecular modelling (using Avagadro software ©) estimated the length and width of SHMP to be 1.5 nm and 0.3 nm respectively. Therefore the change in steric contribution likely relates to a conformational change of the SHMP molecule at the metal oxide-liquid interface. At high pH, the hydroxyl groups of SHMP are dissociated and the molecule becomes increasingly negatively charged (confirmed by the pKa values of SHMP hydroxyl groups), with the molecule attached end-on and perpendicular to the surface. At low pH, the molecule can interact via hydrogen bonding which leads to a slight tilt in the molecule orientation.

The presentation also highlights the effect of ion-concentration (10^{-3} to 10^{-1} M), ion-type (following a Hofmeister series – CsCl, NaCl, and $CaCl_2$), and pH on interaction forces and molecular orientation. The underlying mechanism of the small polyelectrolyte to provide excellent lubricity will be concluded.

Fig. 1. SHMP AdG modelled steric size as a function of pH (in 1 mM NaCl). Inset show the conformational change on SHMP when adopting an end-on and perpendicular orientation at high pH (dominated by electrostatic repulsion between molecules), to a slightly tilted orientation where molecules interact via hydrogen bonding when the pH is below 7.5.

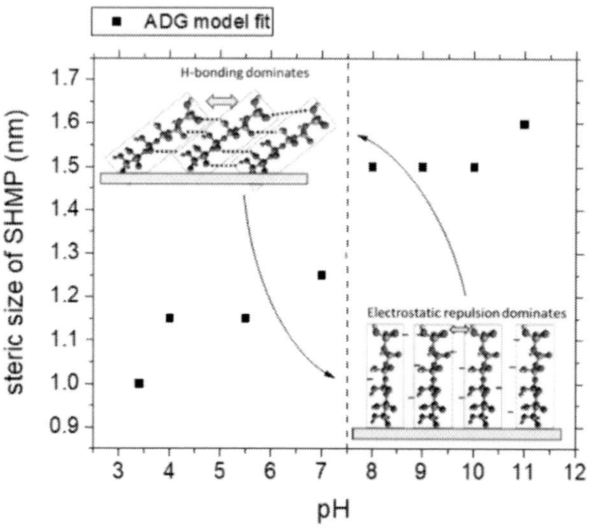

597563 Atomic Layer Deposition of Boron Nitride As a Hydrogen Environmental Barrier Coating

Abstract:

Hydrogen environmental barrier coatings can be used to reduce hydrogen embrittlement of structural materials by reducing diffusion into susceptible substrates. This protection is especially important for hydrogen pipelines and in fields such as nuclear and fuel cell applications where hydrogen is used as a fuel source. The higher the operational temperature, though, the fewer options for material choice. A nanoscale film of the current state-of-the-art material for nuclear applications, tungsten, was previously shown to increase the temperature at which hydrogen reacts with the substrate. However, the increase was still far from the operational temperature of these nuclear applications. In this work, density functional theory (DFT) calculations for energy barriers of atomic hydrogen diffusion were used to predict h-BN as a favorable material for preventing hydrogen diffusion. In addition, the diffusion pathway and the effects of hydrogen in the BN lattice were characterized. To experimentally confirm the effectiveness of BN as a barrier film, particles were coated with nanoscale BN environmental barrier coatings via atomic layer deposition (ALD). The films were then characterized for composition and thickness, and the samples were tested in a high temperature hydrogen environment using differential thermal analysis. Results showed that BN performed as predicted, preventing reaction of hydrogen with the coated sample beyond the maximum temperature of the current state-of-the-art, tungsten.

601486 Predicting Biological Interactions to Monolayer-Protected Gold Nanoparticles Using Molecular Simulation-Derived Descriptors

Abstract:

Gold nanoparticles (GNPs) have attracted significant interest as materials for biomedical applications because they can be detected *in vivo*, deliver drugs, and target specific sites in the body (*e.g.* cancer cells). GNPs are often coated by self-assembled monolayers (SAMs) consisting of alkanethiol ligands with varying backbone and terminal end groups. SAM-protected GNPs are advantageous because of their ease of fabrication with tunable parameters, such as the gold core size, the ligand chain length, and ligand end group chemistry. However, a challenge that inhibits the use of GNPs in biomedical applications is the lack of understanding in how changing these parameters affects the interactions between GNPs and biomolecules, such as proteins or lipid bilayers. The experimental trial-and-error exploration of GNP-biomolecule interactions is cost-prohibitive, motivating the use of alternative techniques to narrow the GNP design space. One promising method to guide GNP design is to relate ligand-specific descriptors to experimental outcomes. For example, recent work has related GNP hydrophobicity – quantified by the octanol-water partition coefficient (logP) of ligand end groups – to phenomena including bilayer adsorption, antimicrobial behavior, and cellular uptake. This approach thus provides some ability to select GNP coatings *a priori* that achieve desired behavior. However, a single ligand end group descriptor is insufficient to characterize changes to GNP surface properties that emerge from variations in GNP size or collective interactions between surface-grafted ligands.

In this work, we use classical atomistic molecular simulations to model SAM-protected GNPs and calculate molecular descriptors capable of characterizing GNP properties while accounting for variations in ligand chemistry, backbone structure, size, and ligand-ligand interactions. We first develop a generalized system preparation workflow that can model a range of SAM-protected GNPs. We then model GNPs in different solvation environments and calculate molecular descriptors that encode information about the GNP surface chemical (*e.g.* hydrophobicity, charge, hydrogen-bonding capabilities, *etc.*) and structural (*e.g.*, eccentricity, ligand fluctuations, solvent-accessible surface area, *etc.*) properties. In particular, we show that GNP hydrophobicity – previously quantified by logP – also depends on SAM, which itself depends on NP core size. We expand the molecular descriptor set to capture GNP-lipid membrane interactions by modeling GNPs in the presence of a lipid bilayer and show that descriptors can predict biological outcomes. The molecular descriptors developed in this work will help predict the effects of GNP functionalization on GNP-biomolecule interactions, opening avenues towards efficient screening of GNPs for selective interactions in the biological environment.

608018 A Cellulosic Shell Installed Onto Nanoparticles Protects Cores from Organic Solvents

Abstract:

We here describe the discovery and characterization of a cellulosic layer on nanoparticle surfaces that protects particle cores from organic solvents. Core-shell nanoparticles 150nm in diameter are formulated using Flash NanoPrecipitation (FNP) and contain a core made of lumefantrine, a model hydrophobic antimalarial drug, and a shell made of hydroxypropyl methylcellulose acetate succinate (HPMCAS), an amphiphilic polymer commonly used as a pharmaceutical excipient. Following FNP, NaOH is added to the nanoparticle suspension, which is then spray dried. During drying, the caustic environment de-esterifies the HPMCAS layer coating the nanoparticles, removing the acetyl and succinyl substitutions from the backbone. The de-esterified HPMCAS – now hydroxypropyl methylcellulose (HPMC) – then vitrifies to form a glassy layer around the NP cores. The HPMC shell layer is impermeable to many organic solvents, and with it in place, less than 10% of the lumefantrine core can be extracted using THF (in which it is readily soluble). Without the NaOH treatment, the THF-soluble HPMCAS shell is dissolved and 100% of the lumefantrine is extracted.

The HPMC shell is characterized by NMR and DSC to confirm its chemical and thermal properties. Necessary conditions for the NaOH treatment and drying are identified, and a method for washing the spray dried powders to remove excess dry NaOH is described. The layer's ability to protect the core is then tested against a variety of common organic solvents. We then describe possible uses for this technique in catalyst and personal care applications, such as temperature-triggered catalyst release or crush-induced fragrance delivery.

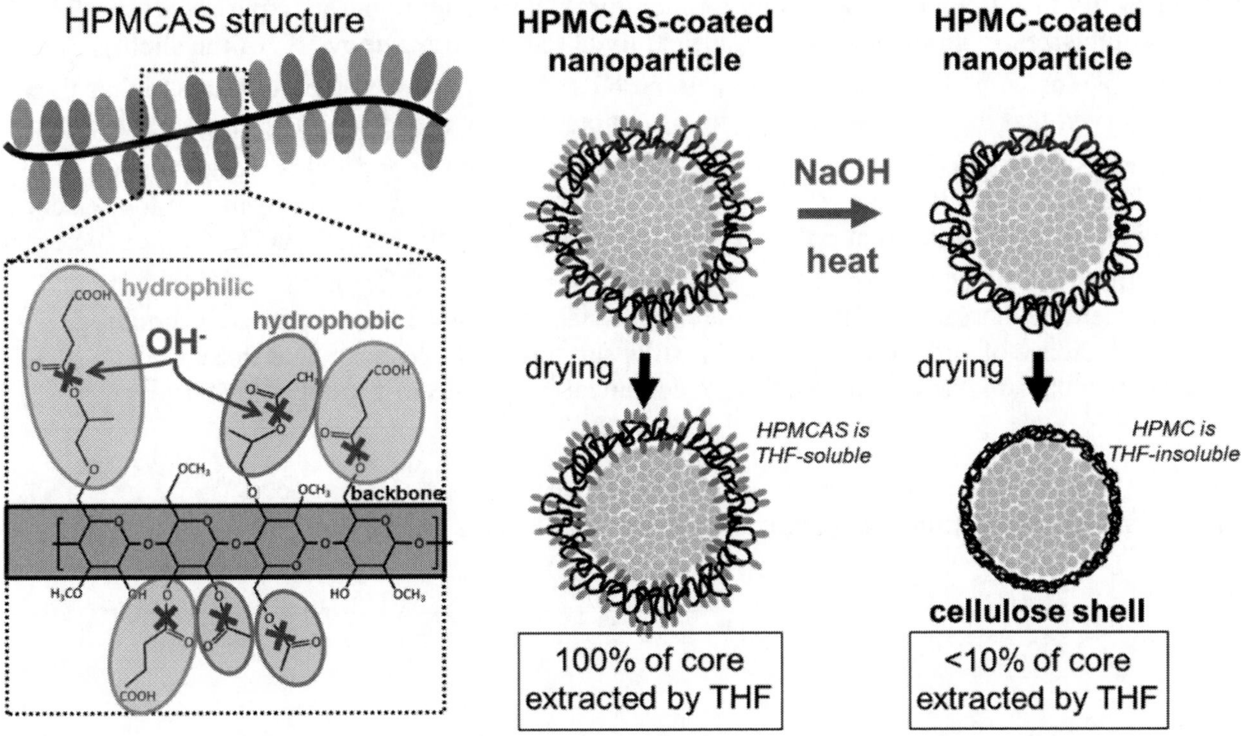

608831 Temperature-Controlled Nanomechanical Shear-Thinning between Polymer Nanoparticle Surfaces during Shear-Induced Aggregation.

Abstract:

Temperature-controlled nanomechanical shear-thinning between polymer nanoparticle surfaces during shear-induced aggregation.

José Francisco Wilson[a, 1](wilsonj@vscht.cz), Miroslav Soos[a] (soosm@vscht.cz).

[a] University of Chemistry and Technology Prague, Technicka 5, 16628 Prague 6, CZECH REPUBLIC

Abstract

In this research, we studied the impact of material fusion as adhesion mechanism on the size and structure of fractal aggregates formed during shear-induced aggregation of fully destabilized NPs. To precisely control particle adhesion, we used nanoparticles (NPs) with a Core-Shell (C-S) structure, where the core is composed of polymethyl methacrylate and the shell is composed of a combination of polymethyl methacrylate and polybutylacrylate. The core acts as a hard-sphere with a high glass transition temperature (T_g) to maintain the structure of the NPs during changes in temperature. On the other hand, the composition of the shell was designed to target a T_g of 50°C , giving the surface of the NPs the capability to soften upon increasing temperature. During aggregation experiments we observed that the size of aggregates grow significantly at temperatures above T_g of the shell, indicating an increase of adhesive force between the NPs. Additionally, at temperatures higher than T_g, it was observed that the increase in adhesion is countered by a reduction of aggregate size when higher shear rates are applied. This suggest that the bond between nanoparticles weakens upon increase of shear rate when the material on the surface experiences a transition from soft to a viscous polymer. This finding indicates that polymer NPs surfaces at temperatures above T_g might experience a local shear thinning effect. To prove this hypothesis, we used DEM simulations and study the interplay between the increase of adhesion due to material softness and the decrease of bonding strength upon increase of shear rate. Obtained results support the hypothesis that these two contradicting mechanisms determine the final mechanical properties or produced material.

[1]

603340 Nano-Coating Stabilized Particulate Photocatalysts for H_2 Production

Abstract:

Photocatalytic nanoparticles that are suspended in water or sprayed onto a panel promise solar-to-fuel energy conversion at scale. The fabrication processes for these photocatalysts to achieve efficient energy conversion should be simpler than solar cells and photoelectrodes. However, the design principles for reaching their performance limit are still elusive. We use few-nanometer thin stabilization coatings to enable an entire class of otherwise unstable particulate semiconductors for coevolution photocatalysis. CdS particles that are sprayed into a film and stabilized with an ALD coating coevolve H_2 at a record rate of 48.5 mmol·h^{-1}·g^{-1} or 2.5 mL H_2·h^{-1}·cm^{-2} under 1-sun solar illumination in ambient air. Moreover, we demonstrate a redox-mediated particulate reactor that separates the H_2 and O_2 products, which is safe. We systematically advance towards the solar-to-fuel efficiency limit of for photocatalysts, instead of trial-and-error that is often practiced in this field. To justify our approach, we will discuss the holistic understanding for the photophysical, electrocatalytic, and transport processes that are coupled at the nanoscale.

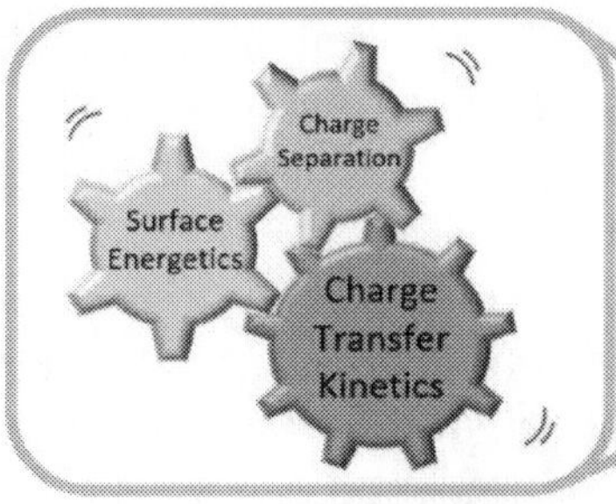

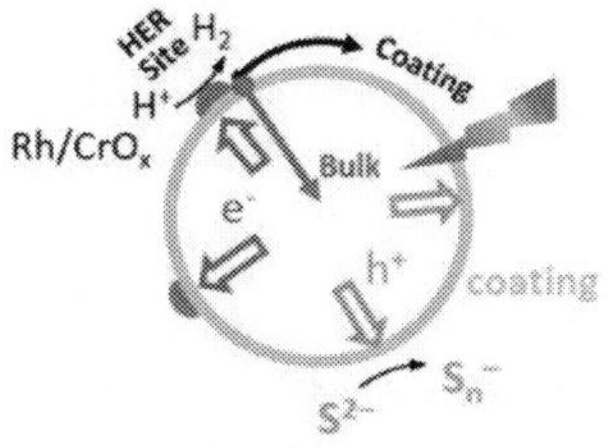

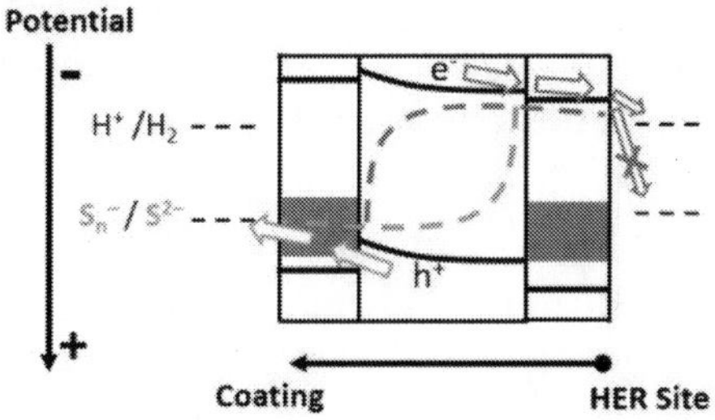

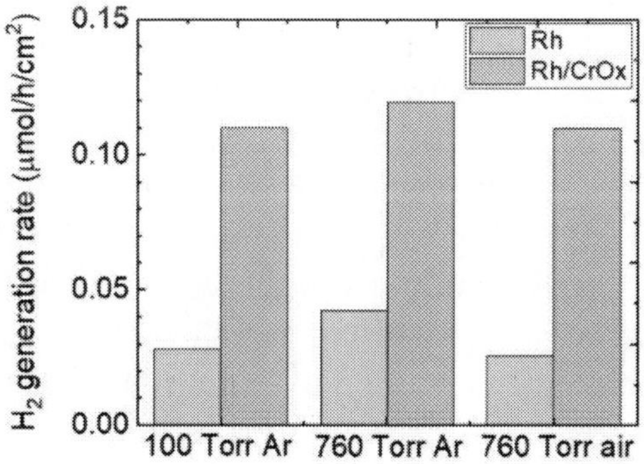

605674 Understanding the Unique Transport, Chelating and Electrochemical Properties of Nanoscale - Based Electrolytes for Sustainable Energy Storage

Abstract:

The recent growth in the worldwide installation of renewable energy has allowed the first steps towards decarbonization of power sector. However, further penetration of renewable energy into the grid is challenged due to the limitations of storage of intermittent renewable electricity. Thus, vast research efforts are on-going to develop electrochemical energy storage systems such as batteries and dense energy carriers. Electrolyte design and selection is a key component of these electrochemical energy systems, mediating transport and solubility of the reactive species to ultimately deliver high energy, current and power density systems. An understanding of stability, electrochemical and physicochemical properties, as well as functionality, is necessary to guide the design of next-generation electrolyte materials for sustainable energy storage systems.

A novel class of nanoparticle organic hybrid materials (NOHMs) has been developed that can serve as breakthrough electrolytes with concentrations of electrochemically active species not achievable in conventional solvents. NOHMs consist of polymer-tethered nanoparticles, where the polymeric canopy is grafted to an inorganic core either ionically or covalently. Due to the polymer tunability, NOHMs are versatile electrolyte media for redox flow batteries and other electrochemical systems, such as electrochemical CO_2 conversion to chemicals and fuels.

NOHMs have been explored as complexing agents of redox active species, increasing their effective concentration in solution. They have been found to effectively chelate copper and zinc redox active species of interest in electrochemical applications, and complexation behavior has been studied via various spectroscopy tools. The electrochemical behavior of NOHMs-complexed copper species has been explored, indicating changes of the overall conversion mechanism through stabilization of the copper (I) state. The degree of chelation of different redox active species has been found to be easily controlled by the solution pH. Tailoring transport properties is also of particular importance in NOHMs-based electrolytes mixtures, as they are challenged by inherently high viscosities, impacting charge transport critical in electrochemical performance. These properties can be tuned by intermolecular interactions between the polymeric canopy and the surrounding fluid. In particular, NOHMs are highly responsive to ionic stimuli, with the addition of even low salt concentrations inducing large reductions in the viscosity of NOHMs-based electrolyte mixtures. Alterations in the degree of polymer swelling and the conformational structure of the NOHMs polymer canopy with ionicity have been probed to explain measured bulk physicochemical properties.

Overall, NOHMs have been found to be electrolyte materials enabling high concentrations of electrochemically active species and with highly tunable transport properties, making them versatile electrolytes for a variety of electrochemical energy storage systems.

609246 Development of Synexomes: Next Generation Lipid-Based Drug Delivery

Abstract:

Introduction

The field of liposomal drug delivery has been burgeoning over the past several decades. However, the efficacy of this drug delivery system is limited by its susceptibility to the human immune system[1,2]. The goal of this research project is to generate completely synthetic lipid nanoparticles (SynExomes) that are produced using a microfluidic system and can persist longer in the human body by resisting phagocytosis, especially upon encountering opsonins and macrophages.

The overall goal of this research project is to generate completely synthetic lipid nanoparticles (SynExomes) that are produced using a microfluidic system and can persist longer in the human body by resisting phagocytosis, especially upon encountering opsonins and macrophages. This goal will be achieved by pursuing three subordinate objectives: i) Modulating the composition of lipid bilayer and conjugating poly-(carboxybetaine) (PCB) to the lipid bilayer to prevent opsonin adsorption on the liposomal surface ii) Surface modification of liposomes by attachment of CD47 protein to delay phagocytosis iii) Determining the impact of the above modifications on SynExome half-life *in vivo*. The proof-of-concept efficacy of this platform will be demonstrated in an infected chronic diabetic wound model where presence of macrophages, opsonins, and inflammation is highly elevated and localized. Here, we describe the first step of the process – synthesis of PCB and its conjugation to the lipid 1,2-distrearoyl-*sn*-glycero-3-phosphoethanolamine (DSPE). We also demonstrate the proof-of-concept functionality of the microfluidic platform for the production of lipid nanoparticles (LNPs).

Methods

DSPE-PCB synthesis involves four distinct steps:

Monomer synthesis: Synthesis of carboxybetaine tertiary butyl ester monomer was performed as described by Cao et al[3]. 2-(dimethylamino)ethyl methacrylate and tert-butyl bromoacetate were dissolved in acetonitrile and reacted under nitrogen protection for 24 h at 50 °C. The monomer was precipitated out of the reaction mixture using excess diethyl ether, dried and stored in a vial under nitrogen protection at -20°C.

Polymer synthesis: Synthesis of NHS-PCB-tBu polymer was performed using a reversible addition-fragmentation chain transfer (RAFT) polymerization reaction[4]. Optimized amounts of the monomer, (RAFT agent), and AIBN were dissolved in *N,N*-dimethylformamide (DMF). The reaction mixture was placed in an ice bath and nitrogen gas was purged into the reaction liquid. After one hour, the purging was stopped, and polymerization reaction was carried out at 70 °C. The unreacted starting material was separated from the polymer product by dialysis.

DSPE-PCB-tBu conjugation: NHS-PCB-tBu polymer and DSPE lipid were dissolved in a mixture of chloroform, DMF and triethylamine and the conjugation reaction was carried out at room temperature for five days[5]. The product fraction containing DSPE-PCB-tBu, unconjugated PCB-tBu and unconjugated DSPE were precipitated out from the remainder of the reaction mix using excess diethyl ether. Acetonitrile extraction was of the desired product (DSPE-PCB-tBu and unconjugated

PCB-tBu) was carried out and the precipitates were dried overnight using hi-vac.

Hydrolysis of DSPE-PCB-tBu: The precipitates containing DSPE-PCB-tBu and unconjugated PCB-tBu were hydrolyzed in trifluoroacetic acid for 4 hours at room temperature. The resulting DSPE-PCB and unconjugated PCB were then precipitated out and unconjugated PCB-tBu was separated from the desired product by diafiltration.

Microfluidic platform evaluation for LNP synthesis: 2.5 mM lipid solution containing 35 mol% cholesterol and 65 mol% DPPC in ethanol was used to make LNPs using a commercially available microfluidic cartridge (NanoAssemblr, Precision Nanosytems). The LNPs were then diluted 1:1 in PBS, dialyzed in PBS to remove ethanol, and filtered using 0.2 µm filter. Dynamic Light Scattering (DLS) and transmission electron microscopy (TEM) were used to determine LNP size and polydispersity.

Significance

The synthetic nature and homogeneity in size of SynExomes achieved through a microfluidic production platform circumvents the issue of variability in size. Use of PCB in the SynExome formulation does not require the use of significant amounts of cholesterol in the bilayer to stabilize it. Therefore, it does not restrict the lipid choices, as in case of PEGylated liposomes. Similarly, a first of its kind CD47 ligand conjugation to lipid through PCB could make these particles more resistant to phagocytosis than conventional liposomes.

References

1. Dams, E. T., P. Laverman, W. J. Oyen, G. Storm, G. L. Scherphof, J. W. van Der Meer, F. H. Corstens, and O. C. Boerman. 2000. "Accelerated Blood Clearance and Altered Biodistribution of Repeated Injections of Sterically Stabilized Liposomes." *The Journal of Pharmacology and Experimental Therapeutics*292 (3): 1071–79. https://www.ncbi.nlm.nih.gov/pubmed/10688625.
2. Wassef, Nabila M., Gary R. Matyas, and Carl R. Alving. 1991. "Complement-Dependent Phagocytosis of Liposomes by Macrophages: Suppressive Effects of 'stealth' Lipids." *Biochemical and Biophysical Research Communications*. https://doi.org/10.1016/s0006-291x(05)80266-9.
3. Shaoyi Jiang, Zhiqiang Cao, Hong Xue, Lei Zhang. 2013. Self-assembled particles from zwitterionic polymers and related methods. USPTO 8617592B2. *US Patent*, filed May 3, 2012, and issued December 31, 2013.
4. Cao, Zhiqiang, Lei Zhang, and Shaoyi Jiang. 2012. "Superhydrophilic Zwitterionic Polymers Stabilize Liposomes." *Langmuir: The ACS Journal of Surfaces and Colloids*28 (31): 11625–32. https://doi.org/10.1021/la302433a.
5. Rajendrakumar, Santhosh Kalash, Ning-Chu Chang, Adityanarayan Mohapatra, Saji Uthaman, Byeong-Il Lee, Wei-Bor Tsai, and In-Kyu Park. 2018. "A Lipophilic IR-780 Dye-Encapsulated Zwitterionic Polymer-Lipid Micellar Nanoparticle for Enhanced Photothermal Therapy and NIR-Based Fluorescence Imaging in a Cervical Tumor Mouse Model." *International Journal of Molecular Sciences*19 (4). https://doi.org/10.3390/ijms19041189.

Results

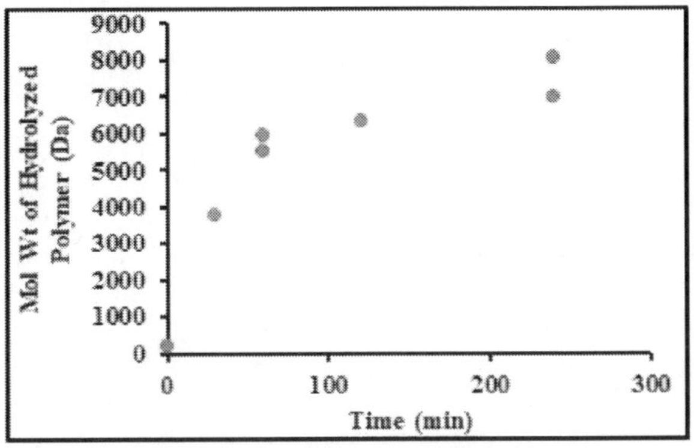

Figure 1: RAFT polymerization reaction kinetics. Reactions run for different time points suggest that 16 minutes of reaction time and 50 minutes of reaction time are enough to synthesize polymers of molecular weight 2000 Da and 5000 Da respectively.

NMR characterization

Monomer: ^{1}H NMR (500 MHz, DEUTERIUM OXIDE) d ppm 1.47 (s, 9 H), 1.90 (s, 3 H), 3.33 (s, 6 H), 3.97 - 4.03 (m, 2 H), 4.30 (s, 2 H), 4.60 - 4.65 (m, 2 H), 5.76 (d, J=1.46 Hz, 1 H), 6.13 (s, 1 H).

Polymer: ^{1}H NMR (500 MHz, DEUTERIUM OXIDE) d ppm 1.50 (br. s., 243 H) 1.61 - 2.17 (m, 39 H) 3.38 (br. s., 147 H) 3.87 - 4.21 (m, 47 H) 4.24 - 4.63 (m, 89 H) 7.45 - 7.57 (m, 2 H) 7.64 - 7.73 (m, 1 H) 7.88 - 8.02 (m, 2 H). Degree of polymerization = 25

DSPE-PCB-tBu: ^{1}H NMR (500 MHz, tfa) d ppm 0.68 - 2.35 (m, 346 H) 3.02 - 3.83 (m, 235 H) 3.89 - 4.97 (m, 151 H) 7.26 - 7.46 (m, 2 H) 7.50 - 7.64 (m, 1 H) 7.79 - 8.05 (m, 2 H) 8.35 (br. s., 12 H).

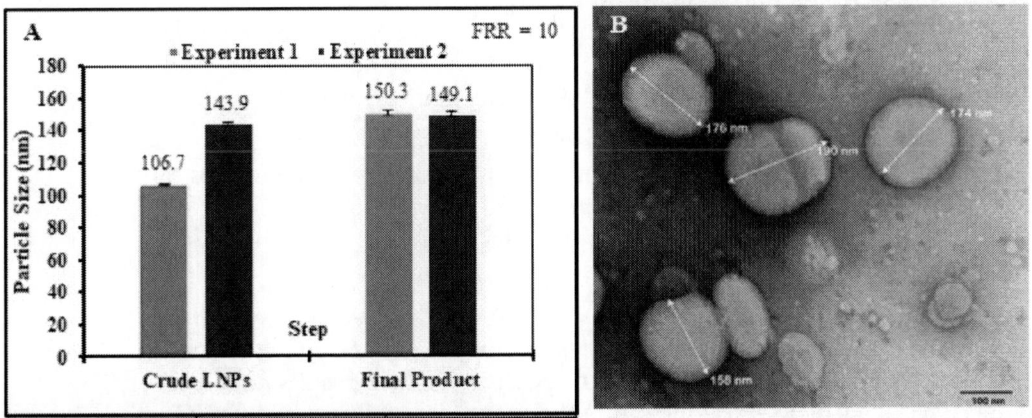

Polydispersity	Experiment 1	Experiment 2
Crude LNPs	0.2	0.2
Final Product	0.2	0.1

Figure 2: (A) Size and polydispersity of lipid nanoparticles obtained using Brookhaven 90 DLS system. Crude LNPs = lipid nanoparticles obtained directly from the microfluidic device. Final Product = lipid nanoparticles post-PES filtration. (B): TEM images of the final product obtained at a magnification 40000X, High Voltage = 80 kV, print magnification: 218000x @ 7.0 in.

610438 Fusogenic Liposomes Loaded with Antibiotics As a Combined Therapy for Fighting Brain Infections

Abstract:

Meningitis is caused by the inflammation of meninges and the in-between fluid filled spaces. Bacterial meningitis, the most common and the most aggressive form of bacterial infection, is mainly caused by Gram-positive bacterium *Streptococcus pneumoniae* and Gram-negative bacterium *Neisseria meningitidis*. Its treatment consists of a combination of antibiotics, that sometimes fail due to their poor permeability through the blood-brain barrier (BBB). Moreover, the emergence of drug-resistant bacteria and their ability to create biofilms have become a global concern, making it increasingly difficult to treat these infections. When bacterial biofilms are formed, the ability of antibiotics to treat these diseases is diminished, leading to ineffective treatment. In severe cases, meningitis can lead to death. Therefore, there is a need to create alternative treatments for bacterial infections of brain, such as bacterial meningitis.

The use of nanotechnology in medicine, especially in drug delivery, is one of the most promising strategies to solve the problem of antibiotic resistance. Here, we report the use of fusogenic (pH-sensitive) liposomes, as drug delivery systems to target the bacteria cell membrane. These liposomes release the antibiotic directly at the site of infection, notably inside the bacterial cell. pH-sensitive liposomes have been previously used to target bacteria due to their ability to interact with the phospholipids present in the bacteria outer membrane, destabilizing it. For the purpose of this study, liposomes were functionalized with a cell-penetrating peptide, TAT (47-57) to increase their permeability through the blood-brain barrier, which is the primary challenge in treating meningitis.

Lipid film rehydration method was used to synthesize the liposomes, followed by their functionalization with TAT (47-57) peptide using wet chemistry. Liposomes were characterized by transmission electron microscopy (TEM) to verify their formation, and size of about 100 nm. These liposomes were loaded with three different antibiotics, methicillin, ampicillin and methicillin, with a loading efficiency of around 68%, 72% and 83% respectively.

Bacterial growth curve studies, and colony forming studies results showed that TAT functionalized liposomes loaded with vancomycin and methicillin were successfully able to kill *Streptococcus pneumoniae*, and methicillin-resistant *Staphylococcus aureus* respectively. These results not only demonstrate the ability of these particles to decrease the minimum inhibitory concentration (MIC) of antibiotic necessary to kill bacteria, but also their ability to combat drug-resistant bacteria.

In vitro cytotoxicity test using fusogenic liposomes against astrocytes and endothelial cells, primary cells that form the BBB, showed no interaction between our nanoparticles and both cell lines, as well as a significant reduction of antibiotic cytotoxicity through encapsulation.

A novel static *in vitro* BBB model was used to determine the ability of these nanoparticles to cross to blood brain barrier. This model consisted of astrocytes seeded at the bottom of a 24-well plate, with inserts containing endothelial cells added to each well. After a set incubation time, endothelial cells were placed inside a separate 24-well plate containing bacteria. Liposomes were added to study their permeability through the endothelial cell barrier to reach the bacteria on the other side of the membrane. Initial studies with these static models showed the ability of our functionalized nanoparticles to not just cross the blood brain barrier, but also significantly decrease the bacteria population on the other site.

Future studies will be focused on a better understanding of the interaction mechanism between our TAT functionalized liposomes and bacterial cells, and on the use of dynamic and more sophisticated BBB models to better mimic the liposome permeability through the BBB.

This work was supported by Department of Chemical Engineering, Northeastern University.

612204 Controlled Loading of Albumin-Drug Pre-Conjugates for Treatment of Pancreatic Cancer

Abstract:

Chemotherapy drugs have been widely used as standard of care to treat cancers for several decades. However, due to their promiscuous cytotoxicity, they can cause adverse events and have narrow dosing regimens, which limit their efficacy. In part to address these challenges, drug delivery systems have been developed to improve the therapeutic index of existing chemotherapy drugs, including drug derivates, micelles, liposomes, and polymeric nanoparticles. However, it remains a challenge to achieve potent delivery of therapeutics in tumors without high levels of accumulation in non-specific tissues. Here, we developed albumin-drug conjugates with controlled loading to harness the intrinsic transport properties of albumin and improve the therapeutic index of current *in situ* albumin-binding prodrugs for significantly improved antitumor efficacy. Here, model drug monomethyl auristatin E (MMAE) was conjugated to Cys34 of albumin via a cathepsin B-sensitive dipeptide linker *ex vivo* to ensure that all drug would be bound specifically to albumin. The resulting albumin-drug conjugate with a drug to albumin ratio (DAR) of 1 (ALDC1) retained the native structure of albumin compared to conjugate with a higher DAR of 3 (ALDC3). ALDC1 exhibited improved drug release and cytotoxicity compared to ALDC3 *in vitro*. ALDC1 exhibited slower plasma clearance and increased drug exposure over time compared to ALDC3 and MMAE prodrug. In single dose studies with MIA PaCa2 xenografts, cohorts treated with ALDC1 had the highest amount of MMAE drug in tumor tissues compared to other treatment arms. After multiple dosing, ALDC1 significantly delayed the tumor growth compared to control treatment arms MMAE, MMAE-linker conjugate and ALDC3. When dosed with the maximum tolerated dose of ALDC1, there was complete eradication of 83.33% of the tumors in the treatment group. ALDC1 also significantly inhibited tumor growth in an immunocompetent syngeneic mouse model that recapitulates the phenotype and clinical features of human pancreatic cancers. In summary, site-specific loading of drug to albumin at 1:1 ratio allowed the conjugate to maintain the native structure of albumin and its intrinsic properties. By conjugating the drug to albumin prior to administration minimized premature cleavage and instability of the drug in plasma and enabled higher drug accumulation in tumors compared to *in situ* albumin-binding prodrugs. This strategy to control drug loading *ex vivo* ensures complete drug binding to the albumin carrier and achieves excellent antitumor efficacy, and it has the potential to greatly improve the outcomes of anticancer therapy.

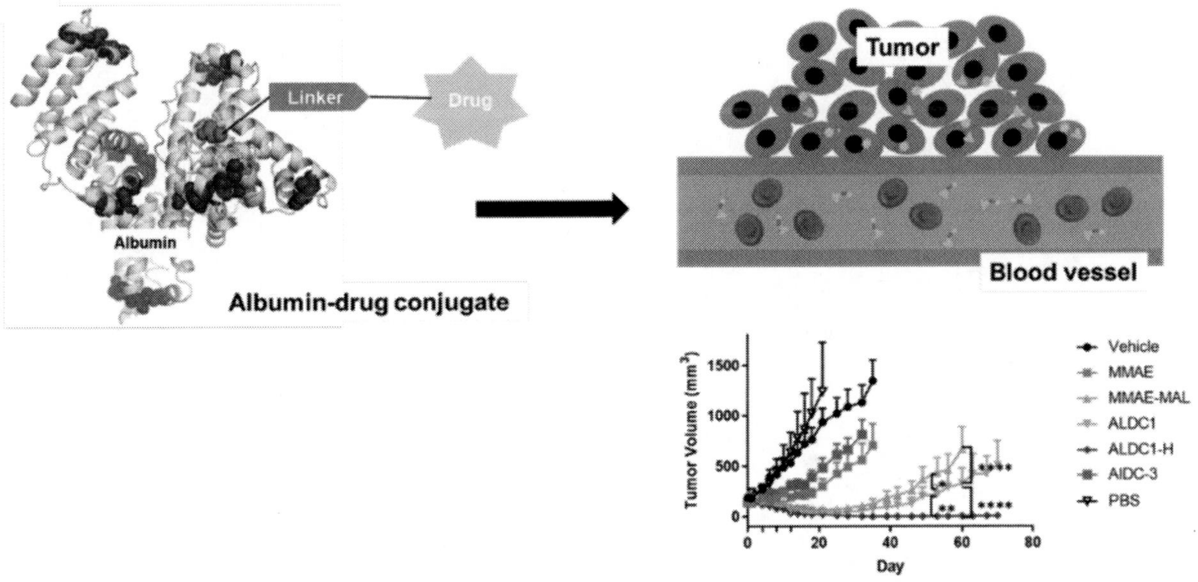

607365 Modified Oxalate-Based Doxorubicin (OX-Dox) Prodrugs for Reactive Oxygen Species (ROS)-Responsive Drug Release

Abstract:

Nanoparticle (NP)-mediated cancer therapeutics have been widely pursued to develop safer and more effective modalities. Although they have made tremendous progress in the fight against cancer, sustained drug release during systemic circulation and controlled release at the tumor site remain as the major challenges. These challenges can be addressed by a controlled release mechanism triggered by internal or external stimuli such as pH, enzyme, reactive oxygen species (ROS), light, ultrasound, and irradiation. Because the ROS level in the cancer cells is much higher than the healthy ones, ROS can be used as an effective stimulus to trigger the drug release.

Here, we developed a few different modified oxalate-based Dox (OX-Dox) prodrugs. After confirming the successful synthesis of these prodrugs using MS and ^{1}H NMR, their release kinetics was studied using different concentrations of H_2O_2 and HRP. Then the MDA-MB-231 and MDA-MB-468 cell lines were chosen to compare and evaluate the efficacy of the synthesized OX-Dox prodrugs using the MTS and clonogenic assays. Also, fluorescent imaging and flow cytometry were employed to compare the *in vitro* cellular uptake of these different OX-Dox prodrugs.

602436 Synthesis, Characterization and Functionalization of Chitosan and Gelatin Type B Nanoparticles with Cell-Penetrating Capabilities

Abstract:

One of the major challenges of modern pharmacology is assuring that the costly developed drugs are able to reach the site of action with significant bioavailability [1]. Often these drugs need to pass across biological barriers and the sequential action of bodily defense mechanisms [2]. Consequently, only a fraction of the intended amount effectively comes in contact with the diseased organ or tissue and manages to penetrate the involved cells. This is particularly challenging for highly polar, highly hydrophobic or charged drugs. To tackle this major issue, a number of strategies have been developed including the conjugation of the active compounds to cell-penetrating peptides [1], and the use of viral and non-viral vectors [3]. Due to issues regarding biosafety, the viral vectors have been criticized and consequently, the non-viral ones have gained a lot more interest over the past few years. A family of non-viral vectors that have demonstrated versatility and ease of handling is the nanomaterials [1]. Due to a worldwide effort, nowadays there is a large arsenal of nanomaterials with different chemistries and topologies that have been successfully tested for drug delivery applications [2]. Some of the evaluated nanomaterials include members of the metallic oxide, carbon-based and polymeric families [4]. In particular, polymeric nanomaterials have demonstrated high efficiency stabilizing and protecting molecules like proteins, peptides or DNA from the degradation of various environmental hazards [1]. Moreover, their synthesis methods are relatively well-established and therefore it is expected to have control oversize, superficial charge, and morphology. These properties are critical in defining the uptake and internalization pathways, efficiency and final fate of the nanocarriers [2].

Even if the nanocarriers are able to penetrate cells, they might be intracellularly trafficked via endocytosis, a process in which the cell engulfs extracellular materials into the cell by invagination of the cellular membrane forming endosomes [5]. Endocytosis can be classified into phagocytosis or pinocytosis. The former is a mechanism for large solutes while the latter is for liquids and small particles. Nanoparticles with size from 250 nm to 3 μm have shown to enter via phagocytosis, while the ones with a size range of 120 nm to 150 nm are internalized via clathrin or caveolin-mediated endocytosis [5]. This specialized route uses ligands that bind to specific receptors in the membrane, which are then engulfed within clathrin-coated vesicles. In any case, the formed endosomes eventually reach lysosomes where degradation enzymes break down the contents. For this reason, to achieve high distribution levels of the drug, the delivered nanocarrier must be able to achieve endosomal escape once inside the cell [5]. Internalization is also a function of surface charge, coating type, and elastic modulus. For instance, soft nanomaterials (e.g., liposomes and hydrogels) tend to have faster and more effective internalization than the hard ones (e.g., metallic oxides) [5].

Studies show that polymeric nanoparticles can be engineered to induce endosomal escape by destabilizing and disrupting endosomal membranes. Moreover, polymers with a buffering capacity can also inhibit the acidification of the endosome, thereby resulting in an influx of chloride counterions and water molecules that increases the pressure and ultimately lyses the endosome. Here, we propose to synthesize two types of polymeric nanoparticles, gelatin type B and chitosan to subsequently functionalize their surfaces with the translocating peptide Buforin II such that the buffering capacity of polymers is synergistically combined with potent endosomal escape moieties [6]. Gelatin B was selected due to its low-cost and superior biodegradability and biocompatibility, while on top of these attributes, chitosan has been reported to have remarkable mucoadhesive and antibacterial properties [7]. Additionally, we will explore the impact of the molecular weight of the polymers for the preparation of chitosan nanoparticles as it has been reported that, depending on this parameter, it is possible to achieve different endosomal escape levels [8].

The gelatin type B nanoparticles were synthesized via the desolvation method where acetone was incorporated as a desolvating agent that is added to an aqueous solution with the material to dehydrate it. As a result, gelatin undergoes a conformational change to form spirals that are crosslinked to form the nanoparticles [9]. Briefly, 200 mg of type B gelatin were dissolved in 10 mL of distilled water followed by dropwise addition of 10 mL of acetone. Two phases were then formed and the supernatant was discarded. The precipitate was redissolved in 10 mL of an aqueous HCl solution (pH 3) under constant stirring (10,000 RPM). Then, an additional 30 mL of acetone and an aqueous solution of 100 µL of 25% (v/v) glutaraldehyde were added to form the nanoparticles. Subsequently, the mixture was left under stirring for 30 minutes to ensure proper crosslinking and the NPs were then centrifuged and thoroughly washed with type I water several times. Finally, the nanoparticles were stored in PBS at 4 °C until further use [10].

The chitosan nanoparticles were synthesized by the ionic gelation method, which is based on the interaction between the positively charged amino groups of chitosan and the negatively charged groups of sodium tripolyphosphate (TPP) [11]. Briefly, Chitosan was dissolved in an aqueous solution of 1.25% (v/v) acetic acid. Then, an aqueous solution was formed at a concentration of 1.25 mg / mL maintained under stirring for 24 hours. After that, the pH was adjusted to 4.7 with a 1N NaOH solution and filtered with a vacuum filter unit equipped with a 0.45 µm PTFE filter. Subsequently, a TPP solution at a concentration of 0.56 mg/mL was added dropwise and kept under stirring for 30 minutes. Then, the nanoparticles were centrifuged and thoroughly washed with type I water several times. Finally, nanoparticles were stored in PBS at 4 °C until further use [11].

Buforin II was conjugated to the nanoparticles by pipetting 50µL of glutaraldehyde to activate the amine groups for one hour. Then, 1 mg of Buforin II was added and left under constant stirring for 48 hours. Excess glutaraldehyde was removed by centrifugation and thoroughly washing with type I water several times.

The obtained nanoparticles were characterized by Fourier transform infrared spectroscopy (FTIR), scanning electron microscope (SEM), confocal microscopy, transmission electron microscopy (TEM). The nanoparticles' size in solution and surface zeta potential were determined with a DLS instrument (Zetasizer Nano). Likewise, to determine the nanoparticles' endosomal escape, the nanoparticles were marked with rhodamine B and visualized via confocal microscopy. The FTIR confirmed the typical bands of the polymers and Buforin II after conjugation. Figure 1a shows the nanoparticles average diameters for gelatin (297±79 nm), chitosan of low molecular weight (182±62 nm), and chitosan high molecular weight (255±52 nm). The zeta potential for both low and high molecular weight chitosan nanoparticles approach neutrality (Figure 1b). This is problematic due to low colloidal stability. In contrast, the gelatin nanoparticles exhibited zeta potential of about -7 mV (Figure 1b). This is close to what is required for colloidal stability, i.e., ±10 mV. Figure 2a shows a TEM micrograph of the low molecular weight chitosan nanoparticles. This image confirms round-shape morphology and sizes in the range of 5-10 nm for individual particles. The SEM micrograph (Figure 2b) shows agglomerates of the low molecular weight chitosan nanoparticles. Similar results were found for the other polymeric nanoparticles.

References

[1] B. Devasier and K. Sanghyo, *Polymer Nanoparticles for Smart Drug Delivery*. Rijeka: InTech, 2014, pp. 257-310.

[2] W. Wang, K. Gaus, R. Tilley and J. Gooding, "The impact of nanoparticle shape on cellular internalisation and transport: what do the different analysis methods tell us?", *Materials Horizons*, vol. 6, no. 8, pp. 1538-1547, 2019.

[3] S. Dizaj, S. Jafari and A. Khosroushahi, "A sight on the current nanoparticle-based gene delivery vectors", *Nanoscale Research Letters*, vol. 9, no. 1, p. 252, 2014.

[4]J. Jeevanandam, A. Barhoum, Y. Chan, A. Dufresne and M. Danquah, "Review on nanoparticles and nanostructured materials: history, sources, toxicity and regulations", *Beilstein Journal of Nanotechnology*, vol. 9, pp. 1050-1074, 2018.

[5] P. Foroozandeh and A. Aziz, "Insight into Cellular Uptake and Intracellular Trafficking of Nanoparticles", *Nanoscale Research Letters*, vol. 13, no. 1, 2018.

[6] D. Elmore, "Insights into buforin II membrane translocation from molecular dynamics simulations", *Peptides*, vol. 38, no. 2, pp. 357-362, 2012.

[7] M. Farshbaf, S. Davaran, A. Zarebkohan, N. Annabi, A. Akbarzadeh and R. Salehi, "Significant role of cationic polymers in drug delivery systems", *Artificial Cells, Nanomedicine, and Biotechnology*, vol. 46, no. 8, pp. 1872-1891, 2017.

[8] S. Smith, L. Selby, A. Johnston and G. Such, "The Endosomal Escape of Nanoparticles: Toward More Efficient Cellular Delivery", *Bioconjugate Chemistry*, vol. 30, no. 2, pp. 263-272, 2018.

[9]R. Donev, *Advances in protein chemistry and structural biology*. Amsterdam: Elsevier, 2015, pp. 169-221.

[10] B. Azimi, P. Nourpanah, M. Rabiee and S. Arbab, "Producing Gelatin Nanoparticles as Delivery System for Bovine Serum Albumin", *Artificial Cells, Nanomedicine, and Biotechnology*, vol. 18, no. 1, pp. 34–40, 2014.

[11] E. Carmona, T. Plaza, G. Recio-Sánchez and J. Parodi, "Generación de un protocolo para la síntesis de nanopartículas de quitosano cargadas con florfenicol a través del método de gelación iónica", *Revista de Investigaciones Veterinarias del Perú*, vol. 29, no. 4, pp. 1195-1202, 2018.

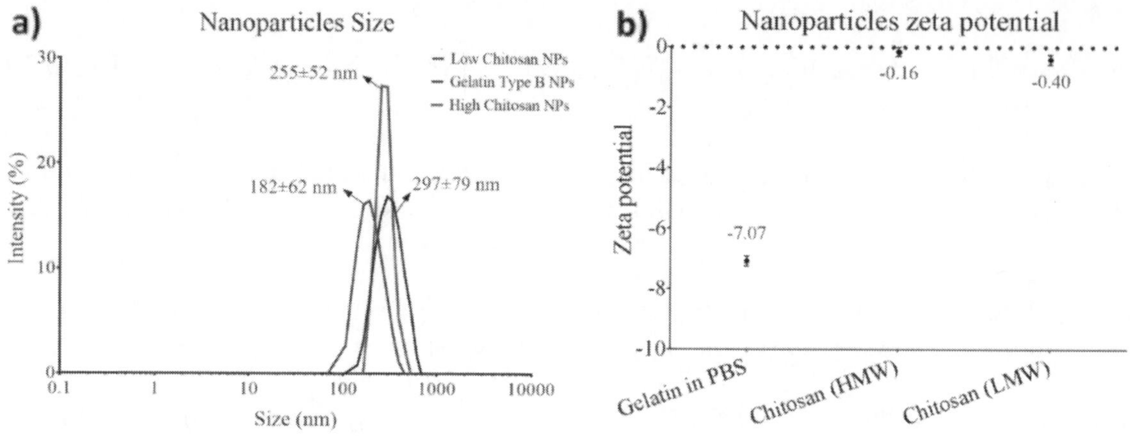

Figure 1. (a) Size distribution of high and low molecular weight chitosan nanoparticles and gelatin nanoparticles, three replicas. (b) Surface zeta potential of high and low molecular weight chitosan nanoparticles in type I water at pH 7 and gelatin nanoparticles in PBS 1X at pH 7.4, experiments were conducted in duplicate.

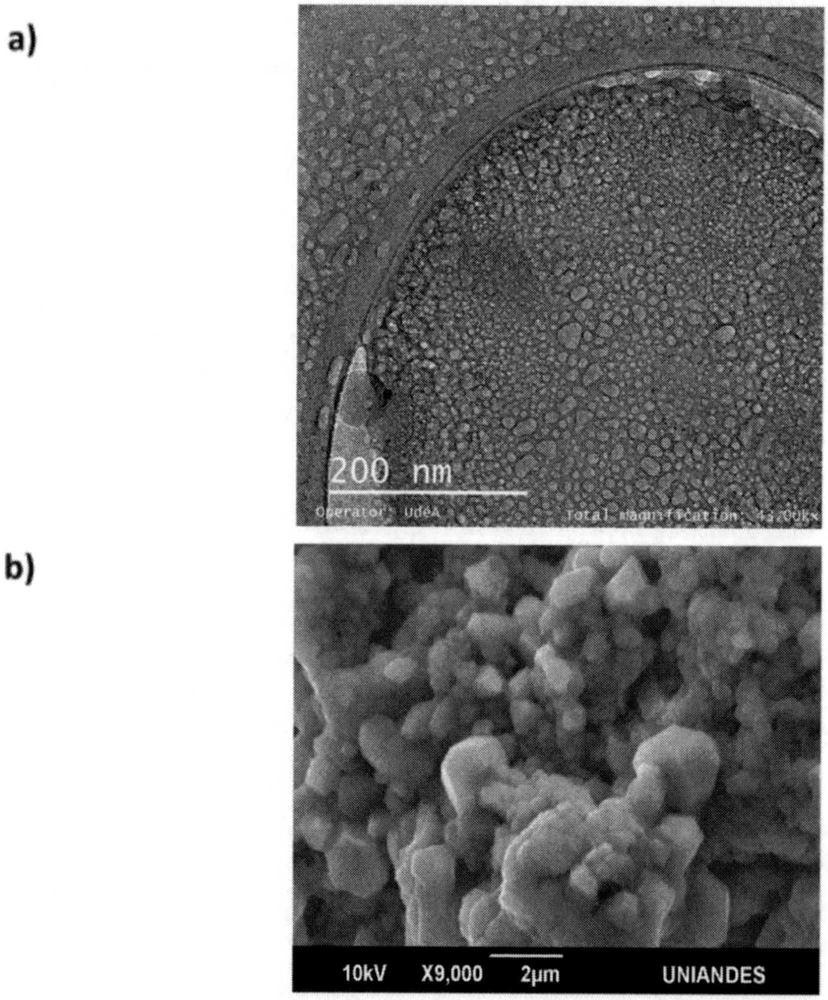

Figure 2. (a) Transmission electron microscopy (TEM) of low molecular weight chitosan nanoparticles. (b) Scanning electron microscopy (SEM) of low molecular weight chitosan nanoparticles.

600519 Temporal Delivery of Polypeptide Nanoparticles for Accelerated Wound Healing and Tissue Repair

Abstract:

Complexity of repair process, low potency of monotherapies and delivery challenges limit the effectiveness of current wound healing treatment strategies, particularly in chronic and diabetic wounds. Growth factors, engineered cells and skin grafts have been investigated for promoting wound healing. However, little attention has been given to the critical role of augmenting early-stage processes in order to set the stage for effective execution of subsequent proliferation and remodeling phases of tissue repair. Here, we demonstrate that temporal delivery of a combination of bioactives that modulate and augment individual phases of healing leads to improved healing outcomes compared to sequential delivery. Histamine, a small-molecule chemoattractant of cells that mediate early stages of tissue repair, was delivered to full-thickness 5 mm excisional splinted wounds in the dorsal skin of immunocompetent BALB/c mice and genetically obese and diabetic db/db mice. Delivery of histamine along with silk fibroin dressing resulted in effective wound closure and improved healed skin strength in both Balb/c and db/db mice compared to Tegaderm dressing alone. Histological and immunohistochemistry analyses showed silk dressing-histamine treatment reduced dermal gap, promoted angiogenesis (CD31+), myofibroblast-mediated wound contraction (aSMA+), and higher TGF-β1 expression which are hallmarks of improved healing outcomes. Growth factors play key roles in different stages of tissue repair, especially in later stages i.e. proliferation, but their availability is limited due to dysfunctional healing response and proteolytic degradation in case of chronic wounds. We have used recombinant methods to generate growth factor nanoparticles (GFNPs) based on a growth factor-elastin-like polypeptide (GF-ELPs) fusion platform. Nanoparticles of basic fibroblast growth factor (bFGF-ELP) and stromal-derived growth factor 1 (SDF1-ELP), with hydrodynamic diameters from 300-400 nm, demonstrated higher stability and bioavailability in wound fluid compared to growth factors alone. Temporally sequential delivery of SDF1-ELP or bFGF-ELP nanoparticles immediately prior to the transition phase following histamine delivery resulted in improved biomechanical and biochemical outcomes of tissue repair compared to simultaneous delivery or individual treatments acting alone. These findings demonstrate that multifactor therapeutics (silk fibroin dressing, exogenous histamine, and GFNPs) are promising for improved outcomes in wound healing and tissue repair, including in diabetic and obese mice, and outperform clinically approved polyurethane wound dressing. A fundamental understanding of tissue repair kinetics and associated temporal delivery of therapeutics can lead to faster healing and improved tissue repair outcomes, thus reducing the propensity for re-opening, progression to intractable phenotypes and surgical site infections, all of which can have remarkable outcomes in the clinic.

600770 Computer-Aided Molecular Design for Chemical and Energy Applications

Abstract:

The energy efficiency and technical difficulties of an industrial chemical process are often determined by the physicochemical properties of materials.[1] It is desirable to design the right structure of chemicals to meet the targeted material properties, such as the adsorption or emission spectrum of light. However, the research and development (R&D) of functional chemicals has been based largely on researcher's experiences and experimental trial-and-error. With the advances in computing power and algorithm, Computer-Aided Molecular Design (CAMD)[2, 3] technique can now serve as an auxiliary methodology to improve the efficiency of R&D. There are two elements in CAMD: First is the methodology for the prediction of molecular properties, including thermodynamic models, quantum mechanical calculation (QM), and molecular simulation etc.; The second part of CAMD is the optimization algorithm that could automatically search molecules with given requirements of molecular properties, including genetic algorithm (GA), simulated annealing algorithm (SA) etc.[4] We have developed a new molecular data structure (MDS)[5] that allows for flexible creation of new chemical structures with GA. In this presentation, we will demonstrate the application of the MDS in three different applications: *(1) Finding new organic solvents of specified value of octanol-water partition function (K_{ow}).*[2, 3] *(2) Finding new organic solvents of specified value of LUMO-HOMO the LUMO-HOMO gap. (3) Finding new ionic liquids (ILs) for CO2 capture.*[6] We show that CAMD could provide new chemicals that have better or comparable performance with well-known/commercialized specialty chemicals.

References

1. Sandler, S., Chemical, Biochemical, and Engineering Thermodynamics. **2006**.
2. Hsu, H. H.; Huang, C. H.; Lin, S. T., Fully Automated Molecular Design with Atomic Resolution for Desired Thermophysical Properties. *Industrial & Engineering Chemistry Research* **2018,** 57, (29), 9683-9692.
3. Achenie, L.; Venkatasubramanian, V.; Gani, R., *Computer-Aided Molecular Design : Theory and Practice.* 1st ed ed.; Elsevier: Netherlands, 2003; Vol. 12.
4. Austin, N. D. Tools for Computer-Aided Molecular and Mixture Design. Dissertation, Carnegie Mellon University, 2017.
5. Hsu, H.-H.; Huang, C.-H.; Lin, S.-T., New Data Structure for Computational Molecular Design with Atomic or Fragment Resolution. *Journal of Chemical Information and Modeling* **2019,** 59, (9), 3703-3713.
6. Wang, J.; Song, Z.; Cheng, H.; Chen, L.; Deng, L.; Qi, Z., Computer-Aided Design of Ionic Liquids as Absorbent for Gas Separation Exemplified by CO2 Capture Cases. *ACS Sustain. Chem. Eng.* **2018,** 6, (9), 12025-12035.

594347 Lignin Carbon Composites As Renewable Components in Electrochemical Devices: Synthesis, Characterization & Modeling

Abstract:

Lignin is a low-cost, renewable resource that, under controlled processing conditions, can be converted into a nanostructured carbon composite with crystalline and amorphous domains. Features of the nanostructure, including the size of the graphitic nanocrystallites and the volume fraction of the crystalline domain, can be tailored through the choice of processing temperature and lignin feedstock. A collaborative experimental and computational research effort has investigated the processing-structure-property-performance relationship for these materials and their application as anodes in lithium-ion batteries. The atomic structure has been investigated using synchrotron x-ray and neutron scattering, complemented by electron microscopy. Interpretation of the scattering patterns has been accomplished via atomistic Molecular Dynamics (MD) simulation and Hierarchical Decomposition of the Radial Distribution Function (HDRDF). The HDRDF approach decomposes the total RDF into a set of atomistic and mesoscale components, which can be rapidly parameterized in terms of structural descriptors including nanocrystallite size and shape, crystalline volume fraction and total density. MD simulations of lithiated and sodiated systems reveal distinct binding mechanisms and explain variations in charging capacity of composites as a function of processing temperature. Electrochemical performance is measured via evaluation of coin cells under a range of operating conditions. The observation that lignin carbon composites demonstrate storage capacities in excess of the theoretical capacity of graphite anodes is explained in terms of a different binding mechanism in these carbon composites.

600606 Efficient Mesoscale Simulation Model for Self-Assembly in Polymeric Systems

Abstract:

The self-assembly in various polymeric systems is the core principle to many of advanced nanotechnologies including energy applications. Success of most of these applications utilizing the self-assembly relies on how well one can adjust, and switch the shape, size and arrangement direction of self-assembled structures. While theoretical and numerical studies have provided valuable insights in understanding the underlying physical principles in the self-assembly of polymeric, often complicated interactions, non-flat free surface, high sensitivity on various system parameters, wide range of length and time scales related with self-assembled structures limit the usages of existing theoretical/numerical models. Moreover, many systems are in metastable states, thus kinetics, not just thermodynamics, plays a key role for the ability of a polymeric material to self-assemble into a desired state. In this presentation, I would like to present our efforts on overcoming the limitation of existing models and extending the scope of numerical approaches to describe experimentally observed microstructures, to predict new mesophases, to examine the suitability of different polymeric systems, as well as to provide the kinetic routes between various microphases in polymeric self-assembly.

606636 Per-Atom Identification: Structure/Dynamics of Supercritical Fluids

Abstract:

At the liquid-gas critical point, the boiling curve terminates and the two phases merge into a single phase of supercritical fluid (SCF). SCF shows an anomalous blend of liquid-like and gas-like traits, which not only leads to industrial applications, but also raises questions on the fundamental physics that governs its exotic physicochemical behaviors. In this work, we answer this question by identifying liquid-like and gas-like molecules coexisting in the SCF, captured by machine learning analysis on simulation data. The deltoid coexistence region encloses the Widom line, and it may therefore be termed the Widom delta. The number fractions are found to follow simplified two-state thermodynamics model, and are closely related to the magnitude of supercritical anomaly which originates from the fluctuation between microstates. Furthermore, we provide evidence that the microstate number fraction is a more fundamental descriptor of supercritical fluid compared to conventional order parameters, e.g. bulk density. These results strongly support modeling the SCF as a mixture of two interchangeable microstates, of which macroscopic properties arise from the microscopic partitioning of the space by molecules.

595769 Computational Study of Perovskite Materials for Highly Efficient Optoelectronic Applications

Abstract:

Due to their superior photophysical properties, metal-halide perovskite (MHP) materials and their nanostructures are receiving great attention. MHPs have demonstrated not only dramatic improvements in photoelectric conversion efficiency, but also high external quantum efficiencies, promising the use of MHPs for solar-cells and light-emitting diodes (LEDs). For successful deployment of MHPs as next-generation optoelectronic applications, however, photophysical properties of MHP materials still need to be further optimized and improved. Toward this end, a fundamental understanding is becoming essential in order to interrelate compositions, structural phases, material stability, and photo-dynamics of MHPs. We thus perform state-of-the-art first-principles-based simulations to unveil complicated relationships among material properties, which will be outlined in this talk. We anticipate that our computational studies will provide theoretical and mechanistic guidelines to develop highly efficient and stable MHP materials for optoelectronic applications.

606608 Tailored Coupling of Biomineralized CdS Quantum Dots to Reduced Graphene Oxide to Realize Ambient Synthesis of a High-Performance Hydrogen Evolution Photocatalyst

Abstract:

Nanocomposite photocatalysts offer a promising route to efficient and clean hydrogen production. However, the multi-step, high temperature, solvent based syntheses typically utilized to prepare these photocatalysts can limit their scalability and sustainability. Bio-synthetic routes for photocatalyst production offer an attractive room temperature and aqueous phase alternative, but often result in poorly performing catalysts. We have developed a method to produce CdS quantum dots (QDs) supported on reduced graphene oxide (rGO) using enzyme-based syntheses and a ligand exchange mediated self-assembly method for a highly efficient hydrogen evolution photocatalyst. All preparation steps are carried out in an aqueous environment at ambient temperature. Size-controlled CdS QDs are prepared through a biomineralization process involving the enzyme-mediated turnover of L-cysteine to H2S in aqueous solutions of Cd-acetate. Graphene oxide is reduced by enzymatically produced H2S to form rGO. Exchange of cysteamine (CA) for the native L-cysteine ligand capping the CdS QDs, as confirmed by zeta potential measurements and ATR-FTIR spectroscopy, enables exploitation of electrostatic interactions to drive assembly of the positively charged CA-capped CdS (CdS/CA) onto negatively charged rGO. UV-vis spectroscopy and TEM imaging offer insight into the CdS loading and spatial distribution on the rGO. The short CA linker molecule allows for efficient charge transfer from CdS to rGO, increasing exciton lifetime and, subsequently, photocatalytic activity. The visible light hydrogen evolution rate of the resulting CdS/CA/rGO photocatalyst was measured at 3300 μmol hr-1 g-1, which is nearly double the rate of the unsupported CdS/CA. This represents, to our knowledge, one of the highest reported rates for a CdS/rGO nanocomposite photocatalyst. Beyond advancing the state-of-the-art in benign, scalable synthesis of high-performance photocatalysts, the ligand exchange strategy developed here offers a potentially generalizable approach for assembling other high-performing, nanocomposite photocatalysts under ambient conditions.

608106 CO_2 Capture and Transport Behaviors of Encapsulated Liquid-like Nanoparticle Organic Hybrid Materials (NOHMs)

Abstract:

The development of hybrid CO_2 capture sorbents has provided a transformative solution to novel water-less and water-lean CO_2 capture solvents that have been challenged with their high viscosities. A number of novel CO_2 capture solvents including ionic liquids (ILs) and nanoparticle organic hybrid materials (NOHMs) have been developed to capture CO_2 from various industrial sources. While NOHMs show thermodynamic advantages for stable CO_2 capture such as strong CO_2 binding energy resulting from functionalized amine groups, great oxidative thermal stability due to silica nanoparticles as a support, low regeneration energy, and negligible vapor pressure, the overall CO_2 capture process is often limited by mass transfer as a result of high viscosity. Therefore, the application of NOHMs in large-scale CO_2 capture requires an innovative design of novel solvent carriers. In order to overcome these issues, we have designed and developed a hybrid CO_2 capture sorbent system based on a unique encapsulation technique, Solvent Impregnated Polymers (SIPs). By immobilizing NOHMs solvents as microemulsions inside a gas permeable polymer (PDMS) matrix, the CO_2 capture rate was significantly improved compared to the bulk NOHMs solvent due to the increase in the gas-liquid interfacial areas and the gas permeability. The governing mechanism of CO_2 capture using encapsulated NOHMs was investigated based on kinetic modeling. The experimental findings suggest that the encapsulation of NOHMs can allow for a flexible design of CO_2 capture units ranging from a fixed bed to a fluidized bed as the sorbents can be treated as a particulate system. The effect of moisture on the CO_2 capture behavior was also investigated as well as the recyclability of the encapsulated NOHMs.

Prepared by LLNL under Contract DE-AC52-07NA27344

AUTHOR INDEX

AIChE
120 Wall Street, FL 23
New York, NY 10005-4020

ISBN 978-1-7138-2315-5